Destiny or Chance Revisited

Planets and their Place in the Cosmos

This exciting tour of our universe explores what we now know about exoplanets and explains the difficulty of finding another Earth-like planet.

Building on the remarkable story of our own solar system from his bestselling book *Destiny or Chance*, Stuart Ross Taylor now takes the reader further, comparing our solar system with the wider universe. How are planets made, and why are they different from stars? Why are exoplanets all different from one another and from our familiar eight planets? What can Earth's nearest neighbors tell us about planetary processes in the whole universe? Why does Earth harbor life?

Beginning with the basic concepts of planet formation and the composition of the universe, the book then summarizes our knowledge of exoplanets, how they compare with our planets, and why some stars have better habitable zones. Further sections provide a detailed study of our solar system, as a basis for understanding exoplanetary systems, and a detailed study of the Earth as our only current example of a habitable planet. The book concludes with a philosophical and historical discussion of topics surrounding planets and the development of life, including why our chances of finding aliens on exoplanets is very low.

This is an engaging and informative read for anyone interested in planetary formation and the exploration of our universe.

STUART ROSS TAYLOR is a trace element geologist and an Emeritus Professor at the Australian National University. His research has covered wide-ranging topics involving trace element geochemistry, from the composition and evolution of the Moon, to tektites and the continental crust of the Earth. He has a D.Sc. from the University of Oxford, is a Foreign Member of the US National Academy of Sciences, and has received the Goldschmidt Medal of the Geochemical Society, the Leonard Medal of the Meteoritical Society, and the Bucher Medal of the American Geophysical Union. Professor Taylor is the author of 240 scientific papers and nine books, including *Planetary Crusts* (with Scott McLennan, 2008), which won the 2010 Mary Ansari Best Reference Award of the Geoscience Information Society. He carried out the initial analysis of the first lunar sample returned to Earth at NASA, Houston in 1969, and Asteroid 5670 is named Rosstaylor in his honor. He is a Companion of the Order of Australia, the highest civilian honor.

Praise for this book:

"We live in an exciting era of discovery. Breakthroughs in solar system exploration, star formation studies, and extra solar planet searches have greatly expanded our horizon of the cosmic root of the Earth and her planetary siblings. A new conceptual paradigm of planetary ubiquity and diversity is firmly taking shape. This book is an encyclopaedic reference of the vast range of intertwining phenomena and processes which compete to shape the paths of planet-making. *Destiny or Chance Revisited* is comprehensive, thorough, and admirably up-to-date. With many intriguing historic anecdotes and vivid analogies, Stuart Ross Taylor lucidly conveys some deep concepts in layman terms without the distraction of intimidating formula or excessive jargon. It is a must-have for all amateurs or professionals who are fascinated by our place in the Universe."
– **Professor Douglas Lin**, *University of California Lick Observatory*

"In his highly-readable style, Ross Taylor describes the most recently discovered members of the solar system family, and the planets found to circle other stars. Anyone interested in planetary formation will be interested by his argument for their formation by chance and the laws of physics, not by destiny or design, and his conclusion that other peopled earths must be extremely rare."
– **Dr. John Wood**, *Senior Scientist (retired), Harvard-Smithsonian Center for Astrophysics*

"This book presents an interesting and novel view of the origin of the Earth and life upon it. It successfully covers the known fact about the main members of our solar system as well as fully covering the recent discoveries concerning other planetary system. It also looks the formation of planets and planetary system, placing the concept within the wider context of stars and galaxies. It is thoroughly readable account, accessible to a wide audience, with complex concepts being explained in an informative way. Readers of all levels will both enjoy and learn from this book."
– **Professor Iwan Williams**, *School of Physics and Astronomy at Queen Mary, University of London*

"Subject: nothing less important than the universe including its history. Author: nothing less than a world-renowned scientist of broad learning with an exceptional gift for exposition. Result: nothing less than a must-read for scientists, philosophers, and anyone interested in learning about some of the most dramatic advances in our understanding of the universe and our place in it."
– **Dr. Michael J. Crowe**, *Professor Emeritus, University of Notre Dame; author of* The Extraterrestrial Life Debate, 1750–1900.

Destiny or Chance Revisited

Planets and their Place in the Cosmos

STUART ROSS TAYLOR

Research School of Earth Sciences
Australian National University
Canberra, Australia

CAMBRIDGE UNIVERSITY PRESS
Cambridge, New York, Melbourne, Madrid, Cape Town,
Singapore, São Paulo, Delhi, Mexico City

Cambridge University Press
The Edinburgh Building, Cambridge CB2 8RU, UK

Published in the United States of America by Cambridge University Press,
New York

www.cambridge.org
Information on this title: www.cambridge.org/9781107016750

First published 2012

Printed and Bound in the United Kingdom by the MPG Books Group

A catalog record for this publication is available from the British Library

Library of Congress Cataloging in Publication data
Taylor, Stuart Ross, 1925–
Destiny or chance revisited : planets and their place in the cosmos /
Stuart Ross Taylor.
pages cm
Includes bibliographical references and index.
ISBN 978-1-107-01675-0 (hardback)
1. Solar system. 2. Extrasolar planets. I. Title.
QB501.T2485 2012
523.2–dc23 2012020244

ISBN 978-1-107-01675-0 Hardback

"UNIQUENESS, IT APPEARS, IS THE COMMON PROPERTY OF PLANETS."

Taylor, S. R. and McLennan, S. M. *Planetary Crusts*, Cambridge University Press, p. 360, 2009.

Contents

Color plates can be found between pages 198 and 199.

Preface

We live in remarkable times replete with technical advances, a consequence of the great intellectual advances of the seventeenth and eighteenth centuries in Europe. *Destiny or Chance* in 1998 looked at the solar system to examine the question whether our planets were likely to be reproduced elsewhere. From the evidence then available, this was judged to be very unlikely, while the possibility of intelligent life resembling *Homo sapiens* [**1**] elsewhere was assessed to be zero. In the succeeding dozen years, major improvements in technology have resulted in the discovery of thousands of exoplanets. Has the situation changed? Yes, in the sense that it has gotten worse. Not only are the exoplanets "Strange New Worlds" as a popular book title has it, but our familiar solar system itself, with its tidy circular orbits, appears to be a rarity. The very architecture of the solar system, familiar to every schoolchild, appears to have arisen through chance collisions and migrations half a millennium after it formed.

Destiny or Chance was written following a close look at our solar system. The numerous planets, satellites, TNOs, asteroids, centaurs and other assorted debris that surround our Sun provided no evidence of design. The resulting array, strange enough when looked at objectively, was clearly the result of a series of chance events. Halfway through writing *Destiny or Chance*, the first exoplanets were discovered. These "Hot Jupiters" were totally unexpected by astronomers, although less surprising to students of the solar system. Lurking in the background is the expectation that something like the Earth, complete with its set of interesting inhabitants, might be discovered.

My excuse for yet another book is to examine this question. I had already concluded in 1992 in the first edition of *Solar System Evolution* that "The planets and satellites are rich in diversity and the difficulty in producing clones of our present solar system makes duplication as unlikely as the possibility of finding an elephant on Mars". Following the publication of *Destiny or Chance,* in an address to the Meteoritical Society, I discussed the problem of the difficulties of making Earth-like planets and came to the conclusion that "the odds of finding 'little green men' elsewhere in the universe decline to zero".

But science moves on. Now the search for Earth-like planets has intensified and numerous examples of Earth-mass planets residing in habitable zones have appeared. Among the reasons for revisiting *Destiny or Chance* is to try to place these in a cosmic perspective. I also wish to investigate the problems of forming an "Earth-like" rather than an "Earth-mass" planet. All have added new hope to the possibility that we are not alone, lost among the incomprehensible spaces that are extended with each new discovery.

I am a physical scientist, not a biologist. The emphasis in this book is on the physical development of habitable Earth-like planets. However major events on the Earth such as worldwide glaciations, the rise of oxygen in the atmosphere, the development of the continental crust, collisions with asteroids and other geological processes have all affected the development of life. The evolution of intelligent life is marked by several improbable events that seem to correlate with these changes and so are addressed here as appropriate.

Meanwhile we have developed a new understanding of the solar system, and of the place of the Earth in it. We understand much about the planets, how they were formed and how they evolved. This enables us to take another look at the idea of "one world or many"? How easy or difficult would it be to make a duplicate of our solar system, or of the Earth, complete with its interesting cargo of inhabitants? Are habitable planets, complete with "little green men (or women)" readily available and common elsewhere? This is one of "humanity's most exciting science projects" as a recent book explains [**2**]. This is a stupendous subject, becoming more complex daily. Here is my attempt to place the current excitement over exoplanets into perspective and to understand our place in the cosmos.

This book is organized into seven chapters, preceded by some notes on the terminology employed. Following the Prologue, I discuss our current understanding of the universe. The next chapter considers how planets are formed. This is followed by a discussion of what we know about the exoplanets. The considerable detail that we now know about the solar system is then summarized. Much of the available evidence about planets is inevitably derived from this source so that there is an unavoidable bias towards our own planets that I have tried, not always successfully, to minimize. The familiar Earth and Moon require yet another chapter. Then I conclude with some perspectives about what it all means.

The sources of various quotations and comments that are identified by numbers in the text (e.g. [8]) are listed by chapter in the Notes at the end of the book. The illustrations are divided into figures, which occur throughout the text, and color plates that can be found in a central section.

Acknowledgments

Much of this book was written while I was a Visiting Fellow in the Department of Earth and Marine Sciences, now part of the Research School of Earth Sciences at the Australian National University. I remain grateful for their hospitality.

I owe a deep debt to many of my scientific colleagues for advice and encouragement that has extended over many years as I have contemplated the problems of planets and our solar system and of our place among all these wonders. The list is far too long to include here. It begins with my schoolteachers of English in New Zealand and concludes with most of the current workers on the problems of planets and the solar system.

I thank Laura Clark, Emma Walker, the production team and particularly Susan Francis of Cambridge University Press for long and continued support in producing this book. Anonymous reviewers for Cambridge University Press made many useful suggestions.

I owe a great deal to Brian Harrold, RSES for much help with computers and who also prepared the color plates. I am also indebted to Dr Judith Caton, ANU, who skillfully turned my rough sketches into polished figures.

I owe much to my patient wife, Dr Noël Taylor for her unstinting support over many years.

Terminology

Planets found outside our solar system are referred to throughout as "exoplanets" as in current usage, rather than as "extra-solar planets". They are generally designated in the literature by the letters b, c, etc. following the name of the star, thus Gliese 581b or HD 149026b. Sometimes the planet is identified by the mission that discovered it; thus Kepler 22b or Corot 3b. I use the term "geology" throughout the text to refer to processes on Earth-mass planets, as it makes for simpler sentences.

Time and distance are particularly difficult to deal with when discussing the universe because both concepts extend far beyond our daily experience. The great contribution of geology to philosophy was to establish the immensity of time or "Deep Time". Comments about intervals of time "as brief as a million years" are common in scientific literature. I avoid the modern scientific convention that refers to the passage of 1 billion years as a gigayear (or the even more appalling abbreviations, Ga or Gyr) because it reduces this stupendous period of time to a trivial level. The origin of the universe as it is currently understood dates back around 13.7 billion years, the time of the Big Bang.

The universe had been around for a long time before the Sun and planets appeared. The time of formation of the solar system has been dated rather precisely. The term ($\mathbf{T_{zero}}$) appears frequently in these pages. It is the earliest reliable date, 4567 million years ago, accurate to within 2 million years that we have in our solar system. What it measures is the earliest formation in our solar system of crystalline material, now preserved as refractory minerals in meteorites. These minerals formed at high temperatures near the early Sun, as dusty grains were recycled through many stages of evaporation and condensation before being spayed out into the solar nebula. Many were trapped in meteorites, where they form a few percent. Some went much farther out and have been found as grains in dust from comets.

They contain minerals formed at high temperatures so these mm-size fragments are thus called "Refractory Inclusions" or "CAIs" (after calcium, Ca and aluminum, Al, two principal components). The significance of the date is that by that time, the Sun had already reached its present mass and was driving gases and volatile elements out from the inner nebula. Perhaps a million years had elapsed since a mass of gas with 1 or 2% of ices and rock had separated from a molecular cloud and started to condense into a star.

Life apparently appeared over 3 billion years ago on this planet. In contrast, it is only 10,000 years since the last ice age ended and the ice that had covered much of Europe and North America, retreated. The whole of recorded civilization is compressed into the past 6000 years.

Distances within the solar system are usually given in terms of the Astronomical Unit. This is the average distance between the centers of the Sun and the Earth, around 150 million kilometers. This useful unit is abbreviated to AU throughout the text. It should not be confused with Au, the chemical symbol for gold, nor with Å, the Ångstrom (10^{-8} cm), another useful measure that is about the size of an atom. The planets extend out to the orbit of Neptune at about 30 AU. The outer boundary of the solar system is at the edge of a spherical cloud of comets that extends to about 50,000 AU. Light takes almost a year to reach us from that distant region.

All these immense expanses are trivial on an astronomical scale. For these vast regions, the distance travelled by light in a year, the so-called "light year" which is about 63,000 AU, now becomes a more useful measure. The nearest star is about 4 light years away. A more frequently used unit in astronomy is the parsec, 3.26 light years.

One of the most striking features of our planetary system is that the planets orbit close to a plane. This is defined by the orbit of the Earth around the Sun and is generally called the plane of the ecliptic. The tilt or obliquity of the planets refers to how far the spin axis of the individual planet is tilted relative to this plane. Thus the tilt of the axis of rotation of the Earth varies around 23°, a feature that provides us with the seasons that we so greatly admire, as the northern or southern hemispheres receive more or less sunlight.

Two other terms dealing with the orbits of planets need to be mentioned. These are the inclination and the eccentricity of the orbits. Inclination is the angle that the orbit of the planet, asteroid, comet or whatever makes to the plane (ecliptic) in which the Earth rotates around the Sun. Except for Mercury with a 7° tilt, our planets have inclinations within a few degrees of the plane of the ecliptic. Other bodies like Pluto (17°) and many of its companions in the Kuiper Belt, along with many comets, have high inclinations. Those exoplanets that are observed to transit, or cross in front of their parent star likely have low inclinations.

How far the orbit of a planet departs from a perfect circle ($e = 0$) and becomes oval or elliptical is measured by its eccentricity. Kepler established that the orbits of our planets are elliptical, although they do not in fact deviate very far from circular. Many bodies in the Kuiper Belt, including Pluto ($e = 0.25$), have eccentric orbits. An extreme case in our system is the TNO Sedna ($e = 0.85$) which has an extremely eccentric orbit that takes it from 76 AU at its closest approach to the Sun, out to 960 AU at the farthest point. Many exoplanets have similar highly eccentric orbits.

Another important feature of orbits, resonance, needs to appear here, as it is significant in many aspects of planetary dynamics, for example, as in the strange orbit of Pluto. This icy dwarf along with companions orbits the Sun two times for every three orbits of Neptune and so it is referred to as being in a 2:3 resonance with Neptune. The importance of these simple whole number resonances is that, in each orbit, the two bodies return to exactly the same relative positions in space. So their minute gravitational interactions accumulate, rather than being evened out. Sometimes this results in stability, as with Pluto, but other resonances clear out gaps, as in the 2:1 resonance of asteroids with Jupiter.

Bodies in close orbits, like our Moon, may become tidally locked, always presenting one face to their companion. If this happens to be a planet near a star, unfortunate consequences follow with one side too hot and the other too cold.

Another convention I have to mention deals with temperature scales. In addition to the familiar Centigrade (or Celsius) and Fahrenheit scales, the Kelvin scale is commonly used in science. It uses the same intervals as the Centigrade scale but is expressed simply as K (not to be confused with the same symbol used to indicate 1000, nor with the element symbol for potassium (aka kalium)). Absolute zero on the Kelvin scale is the temperature at which all motion of molecules ceases. It is 273° below zero on the Centigrade scale which is set by the freezing point of water. One of the coldest places in the solar system is the surface of Triton, the satellite of Neptune, which has a surface temperature of a mere 38 K.

Percentages are the unit that is commonly used in talking about the abundances of the common chemical elements. Another convenient unit is "parts per million" (one part in 10^6), usually abbreviated to "ppm". One percent (one part per hundred) is 10,000 ppm. Ppm is a useful unit for comparing the abundances of trace elements, mainly because it enables us to use small numbers and so avoid long strings of zeros that easily allow errors to creep in. For example, the concentration of uranium in the crust of the Earth is usually referred to as 3 ppm (rather than 0.0003%), while the total amount of water in the Earth amounts to about 500 ppm, less easily confused than 0.05%.

Parts per billion (or ppb, one part in 10^9) is employed for abundances 1000 times lower than ppm. Thus the amount of the element iridium in the Earth's crust is only one tenth of a ppb. In contrast, this element is 5000 times more abundant in meteorites, where it is present at 500 ppb or 0.5 ppm. Because of this extreme difference, concentrations of iridium in the crust as low as 10 ppb are 100 times more than the average and so are commonly signatures of the impact of a meteorite on the Earth. The most famous example is that of the asteroid collision that destroyed the dinosaurs. This event left a measurable spike of iridium at the Cretaceous–Tertiary boundary around the globe from Denmark to New Zealand.

In the astronomical world, all elements heavier than helium, including such distinct elements as chlorine, nitrogen, oxygen and sulfur and the rest of the Periodic Table, are referred to as "metals". This has become standard usage for astronomers, who have much more serious problems to think about than the details of chemistry. The easily measured ratio of iron to hydrogen (Fe/H) is commonly used as a measure of the "metallicity" of the star, although it is only an approximation as low-mass elements such as oxygen form by different nuclear processes. The convention has arisen because iron is a common and readily measured element in stars and so proxies for everything else in the Periodic Table. The story of the formation of the chemical elements, worked out through a combination of nuclear physics, astrophysics and astronomy in the 1950s, is one of the great triumphs of human understanding of the universe, which can only be mentioned here in passing (See Burbidge, E. M., Burbidge, G. R., Fowler, W. A., and Hoyle, F. Synthesis of the Elements in Stars, *Reviews of Modern Physics*, Vol. **29**, 547–650, 1957).

The point of emphasizing the use of the term "metals" here is that planets commonly form around metal-rich stars. Metal-rich is perhaps a misnomer as the metal content of stars ranges from near zero with 10,000 times less metal than our Sun to a maximum of 3 or 4%, our Sun containing about 1.4%.

Throughout the book I refer to the Edgeworth–Kuiper Belt of icy planetesimals beyond Neptune as the Kuiper Belt although this is historically incorrect. However it has become common usage and certainly makes for simpler sentences. Icy bodies in the outer reaches of the solar system are referred to as Kuiper Belt Objects

(KBOs) if they inhabit the Kuiper Belt. A more general term, Trans-Neptunian Objects (TNOs) includes both the Kuiper Belt bodies and those further out in the Opik–Oort Cloud, here again referred to as the Oort Cloud.

The famous astronomical classification of stars, dating from 1920, of OBAFGKM (with newly added L, T and Y classes) covers the range from very hot giant stars (O class) to the cooler red dwarfs (M class) and brown dwarfs (L,T,Y). The Sun is a G class star and F, G and K stars are thought most likely to harbor planets on which life might develop (Figure 1).

Abbreviations and symbols

AA	Astronomy and Astrophysics
AAR	Astronomy and Astrophysics Review
Al	Aluminum
Ag	Silver
ApJ	Astrophysical Journal
ApJ: L.	Astrophysical Journal Letters
ASP	Astronomical Society of the Pacific
Au	Gold
AU	Astronomical unit
Ba	Barium
BJOG	An International Journal of Obstetrics and Gynaecology
Ca	Calcium
CAI	High temperature (calcium–aluminum) mineral inclusions in meteorites
CH_4	Methane
CO	Carbon monoxide
CO_2	Carbon dioxide
Cu	Copper
D/H	Deuterium/hydrogen ratio
ESA	European Space Agency
ESO	European Southern Observatory
Eu	Europium
Fe	Iron
GRL	Geophysical Research Letters
HARPS	High Accuracy Radial Velocity Planet Searcher
HD	Henry Draper, who funded a star catalog
Hf	Hafnium
HR	Hertzsprung–Russell

IAU	International Astronomical Union
Ir	Iridium
K	Potassium
K/U	Potassium–uranium ratio
K-T	Cretaceous–Tertiary
KBO	Kuiper Belt Object
LHB	Late Heavy Bombardment of the Moon, 4000 million years ago
M_E	Planetary mass relative to Earth
MER	Mars Exploration Rover
Mg	Magnesium
M_{Jup}	Planetary mass relative to Jupiter
Mo	Molybdenum
MORB	Mid-Ocean Ridge Basalt
NASA	National Aeronautics and Space Agency
NEA	Near-Earth Asteroid
NEO	Near-Earth Object
Pb	Lead
Pt	Platinum
QJRAS	Quarterly Journal of the Royal Astronomical Society
SETI	Search for extraterrestrial intelligence
Sr	Strontium
REE	Rare Earth elements
Sn	Tin
Th	Thorium
TNO	Trans-Neptunian Object
T_{zero}	4567 million years ago, derived from the earliest dated minerals in our solar system
U	Uranium
USGS	United States Geological Survey
W	Tungsten
Zn	Zinc
Zr	Zirconium

1 Prologue

WHAT IS A PLANET?

> "When we think of a planet, our first conception is a body like Earth with an atmosphere, continents and oceans" [1].

This question is less important than one might suppose, given the uproar about the status of Pluto. Although labels are useful, trying to define a planet runs into the philosophical difficulty of attempting to classify any set of randomly assembled products. A bewildering array of objects form in the nebular disks around stars. These items include in our system, dust, asteroids, Trojans, Centaurs, comets, TNOs, our eight planets from tiny Mercury to mighty Jupiter and their 160 satellites. All differ from one another in some salient manner. A rational view would merely define our planetary system as having four planets (the gas and ice giants) with some assorted rocky rubble sunwards and icy rubble beyond. The significant question is how did they form and evolve, not what pigeonholes this variety of objects can be forced into. The strange varieties of exoplanets and brown dwarfs have added much extra complexity [2].

The views of astronomers and planetary specialists on what should constitute a planet have varied widely, but these often reveal as much about the commentator as the problem. Planets are not something to be tacked on to the bottom corner of the Hertzsprung–Russell diagram. If we use the three physical properties of orbital characteristics, mass and roundness, this leads to a total of 24 planets that includes many satellites. This classification is too broad to be scientifically or even culturally useful. As the New York Times has remarked "too many planets numbs the mind".

The definition of planet by the International Astronomical Union in 2006 ordained that in our solar system, a planet

(1) is in orbit around the Sun,
(2) has sufficient mass to assume a nearly round shape,
(3) has cleared the neighborhood around its orbit.

Any body, except satellites, that meets only the first two of these criteria is classified as a "dwarf planet". So the decision of the IAU that there are eight major planets and five dwarf planets in our planetary system seems an appropriate compromise. A further category of "minor planet" includes the asteroids, Trojans, NEAs (Apollos, Atens and Amors), Centaurs, comets, TNOs and KBOs. Brian Marsden made the useful observation that "it has rarely been scientifically useful to use the word [planet] without some qualification" and terms such as ice giants or terrestrial planets will always be needed. At least in our solar system, it is useful to recall the wise words of Confucius "The beginning of wisdom is to call things by their right names".

The problem with exoplanets is only apparently less acute, as we currently detect only planetary-size bodies. Nevertheless a problem soon arose of how to distinguish planets from brown dwarfs. By convention these bodies lie between 13 Jupiter masses, the lower limit for deuterium burning, and 80 Jupiter masses, above which hydrogen fusion becomes possible, so enabling red dwarf stars to form.

But confusingly, free floating objects down to about 3 Jupiter masses have been found and so are termed "sub-brown dwarfs".

Although the upper limit for giant planets was originally set at 12 or 13 Jupiter masses, strange new worlds continue to appear to confound those who wish for a tidy classification. To cloud the issue further, Corot 3b, a planet with over 22 Jupiter masses, has been found in a 4-day orbit. Other inhabitants of the brown dwarf desert have appeared. So the upper limit of 12 or 13 Jupiter masses for planets has now been changed to 25. Perhaps we need to retreat to a definition that planets are objects that form in disks around stars and arise by a different process (bottom-up) than stars and brown dwarfs that form in gas clouds (top-down). But the complexity of the problem of classification merely points up the message of this book: chaotic events rule the planetary world [2].

WHY CAN'T A PLANET BE MORE LIKE A STAR?

> "The brown dwarf desert . . . strongly suggests that the vast majority of exoplanets formed via a mechanism different from that of stars" [3].

Planets are individuals formed by stochastic processes. They resist generalizations and being placed into pigeonholes. The discovery of over several thousand exoplanets orbiting stars other than the Sun has brought the question of planetary origin and evolution into sharp focus, following from 40 years of exploration of our own solar system.

The detailed study of planets is in fact a very late development in science. It has required the earlier development of many other disciplines. The intellectual leap from the biblical chronology into deep time was mostly due to James Hutton, a member of the Scottish Enlightenment, in the late eighteenth century. The Hertzsprung–Russell diagram, fundamental to astronomy and astrophysics, dates from 1913. The robust OBAFGKM classification of stars (with recent L, T and Y additions for brown dwarfs that spoil the famous mnemonic) also appeared nearly a century ago. Al Cameron finally clarified the origin of the Moon in 1984, while the problem of the origin and evolution of the planets in the solar system is only now slowly coming into focus.

A major problem in trying to understand planets is that, unlike stars, they are individuals that refuse to be placed into a tidy classification. While stars are relatively uniform in composition (except for metal contents from near zero to 4%) and differ mostly in mass, planets are assembled in the late stages of star formation from the leftover debris in nebular disks, and so resemble the products of a junkyard.

Jupiter is not simply a failed star and this illustrates the dilemma. Like the lament in *My Fair Lady*, "Why can't a woman

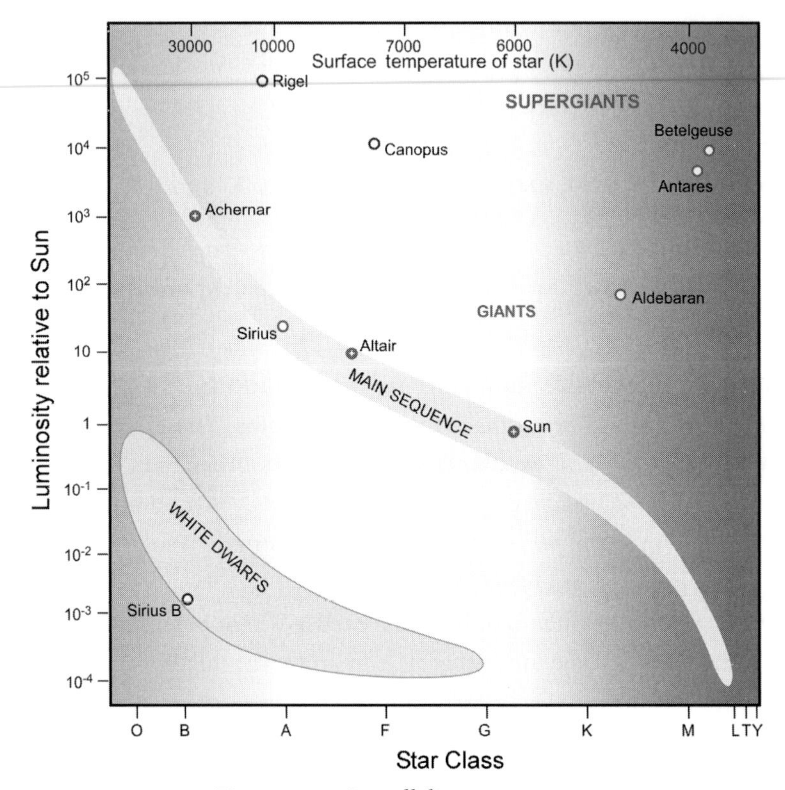

FIGURE 1 Hertzsprung–Russell diagram

The famous Hertzsprung–Russell diagram plots the surface temperature of a star (or class of star) against the luminosity of the star relative to the Sun on a logarithmic scale. The star classes include L, T and Y classes (brown dwarfs) that lurk in the bottom-right corner, as well as the classical OBAFGKM classes. Some well-known stars are identified. See also color plates section.

be more like a man", life would be simpler if planets were more like stars. But there is no equivalent of the Hertzsprung–Russell diagram (Figure 1) that might define the basic parameters for planets, nor much sign of one appearing and that's what makes the subject difficult. We are essentially dealing with individuals. It is even difficult to arrive at a satisfactory definition of a planet as noted; witness the furor over the status of Pluto, which is an eccentric dwarf when placed among the planets, but is one of the largest ice dwarfs in the Kuiper Belt in its own right.

In our solar system, we have eight planets, all different from one another in significant ways. This of course is a celebrated example of the statistics of one, but here we also have four major satellite systems and over 160 satellites. But the regular satellites of the giant planets could just as well belong to different planetary systems. If satellite systems are an inevitable by-product of building planets, the end products are startlingly diverse. Our limited sampling of the exoplanets shows few similarities to what we see in our own system in terms of mass and spacing of planets, while many are in highly eccentric orbits, unlike our tidy near-circular orbits.

So there is a philosophical difference between dealing with stars and with planets that requires a new type of scientist with a distinct mindset, somewhere between the mathematical sophistication of astrophysics and the geological sciences, whose detective-like approach Sherlock Holmes would recognize. As Clemenceau famously remarked: "War is too important to be left to the Generals", and the study of planets is too significant to be left to any one specialist group, either geologists ensnared by the properties of this unique planet, or astrophysicists beguiled by the physics of star formation.

The problem is typified at present by the two competing theories for the formation of the exoplanets: top-down by condensation from the nebula, or bottom-up, by gravitational collapse of gas around earlier-formed cores. Planets formed by the first process might be expected to have a similar composition to that of the Sun, but the giant planets in our own system do not have solar compositions and appear to have formed by the second mechanism. Among other evidence for the latter model is the very existence of the ice giants Uranus and Neptune, 14 and 17 Earth masses respectively, that are mostly composed of rock and ice (or metals) with only one or two Earth masses of gas. They thus constitute analogs for the cores of Jupiter and Saturn.

The observation that the extra-solar gas giant planets preferentially form around metal-rich stars implies that metal-rich cores are also needed to build gas giants elsewhere, at least to followers of William of Occam.

In contrast with the gas giants, the existence of Earth-like exoplanets cannot be addressed directly in the absence of current examples, and so resembles astrobiology. Information from our own system reveals the obvious requirements for metals, orbits of low eccentricity and the avoidance of giant planet migration into the inner nebula that would have cannibalized the terrestrial planets. We also know from their diversity of chemical composition that the formation of our rocky or terrestrial planets (Mercury, Venus, Earth and Mars) was essentially stochastic.

The formation of the Moon provides some clues. Although several people, notably Bill Hartmann, were beginning to understand that collisions were involved with the origin of the Moon, it was Al Cameron who produced the smoking gun in the form of angular momentum conservation. One had to hit the Earth with something the size of Mars, to account for both the rapid spin of the Earth–Moon system and to splash off the rocky mantle of the impactor, to form the low density Moon. But even if the Moon is a special case, it provides the evidence from its pockmarked face of the validity of the planetesimal hypothesis (the accretion of the terrestrial planets from a multitude of smaller rocky bodies) and so tells us, as do the meteorites, much about early events in the inner solar system.

But making these planets in our system is not just a matter of accumulating the rocky fraction from the primordial nebula. The inner nebula, from which the rocky planets formed, was also depleted in elements whose sole common property is volatility. Ironically, this depletion, which has occurred out to several AU, also depletes the habitable zone around the Sun in biologically important elements, such as C, N, P and K. The inner nebula was also bone-dry, as shown by the anhydrous nature of the primary minerals of meteorites (olivine, pyroxene and plagioclase), so that the splash (500 ppm) of water in our planet had to come in later from the neighborhood of Jupiter.

Even when nature gets around to building two similar planets, it finished up with the Earth and Venus. These twins, unlike Mars and Mercury, are close in mass, density, bulk composition and in abundance of the heat-producing elements (potassium, uranium and thorium), but the geological histories of these "twin" planets have been wildly different. Thus plate tectonics, which has the useful property of both building continents and forming ore deposits useful for advanced civilizations and so enabling this discussion to take place, appears to be unique to the Earth.

Venus, in contrast, is a one-plate planet and appears to undergo planetary-wide resurfacing with basalt perhaps every billion years. What is the difference due to? The short answer is water, recalling the aphorism "no water, no granites, no oceans, no continents". As the study of Venus shows, similarity does not mean identity.

So the problem of forming planets elsewhere would seem to depend on repetition in detail of the essentially random processes of planetary accretion and subsequent geological evolution that have characterized the formation of planets in our solar system. As it is very difficult here to form a clone of the Earth, it would seem difficult to make Earth-like habitable planets elsewhere. One must recall the difference between necessary and sufficient conditions for the emergence of life. Much more is needed than metals, orbits of low eccentricity within the habitable zone, water and plate tectonics. The concept that Earth-like planets are rare is not a conclusion that one might wish for, but like much in science, this is what can be read from the observations.

So the study of planets remains in its infancy, beset by the random nature of planet-forming processes and of their subsequent evolution, with the problems requiring the collaboration of many disparate scientific disciplines. This has resulted in a torrent of literature, heavy with speculation. This deluge is reminiscent in many ways of the flood of papers, now mercifully forgotten except by those old enough to remember, that appeared in the 1960s, prior to the Apollo landings on the Moon. It is curious to record that at that time, nearly every possible composition was suggested for the surface of the Moon, except what was revealed by the returned lunar samples.

ANCIENT PERSPECTIVES

It has taken us a long time to discover where we are. Primitive tribes living in remote jungle valleys have often been astonished to discover that the Earth extends far beyond their limited horizon and that they are not its only inhabitants. Before Copernicus, it was generally believed in the civilized world that the Earth was the center of the universe. However, it has slowly been realized that we live in a bigger arena. When you look up at the sky on a dark night in the country, the most striking feature, when the Moon is down, is the glowing band of stars, referred to as the Milky Way, a term first used in English literature by Geoffrey Chaucer (1342–1400) in 1384. This glowing band of stars spreading across the heavens is an edge-on view of our galaxy from the inside.

Although there is a place for the Milky Way in most mythologies, before recent times only Kant seems to have realized what we were looking at. From a nearby galaxy, a few hundred thousand light years away, the magnificent spiral structure that is obscured by our edge-on view would be revealed in its entire splendor. But even this enormous spiral system is only a tiny portion of the universe. Each new telescope reveals a larger universe than our imagination had conceived. Like travelers lost in a desert wasteland, we urgently seek for signs that we are not alone.

THE VIEW BEFORE COPERNICUS

Here I summarize what the ancients made of the world in which they found themselves, as civilization slowly arose following the melting of the great ice sheets about 12,000 years ago. Many of our present notions were formulated in the great flowering of civilization in Greece and Rome. Indeed it has been suggested that the rise of scientific enquiry was not an inevitable development, but an accident, possible only in societies such as ancient Greece and Western Europe in the seventeenth and eighteenth centuries [4].

Babylonian and Greek astronomers observed the strange motion of the planets against the fixed positions of the stars. In this manner, they became aware that there were two classes of heavenly objects in addition to the Sun and the Moon. The term "planet" is derived from the Greek word meaning "wanderer". Curiously there is no mention of planets as distinct from stars in the Old Testament of the Bible. The "Star of Bethlehem", famously depicted on Christmas cards, is recorded only in the gospel according to St. Matthew in the Authorized Version of the New Testament. The most credible explanation, if it is more than a myth, is that it was a conjunction of Jupiter and Saturn that occurred three times during 7 BCE [6].

It is curious that, although the ancient astronomers devoted much study to the movements of the planets, they did not spend much time considering the origin of the solar system. The planets were mostly distinguished from the other heavenly bodies by their wanderings. The whole question of origins seems to have been the province, not of the astronomers, but of the philosophers. There was no shortage of these, or of their ideas.

Some astronomers, however, took up the challenge. Among them was Anaxagoras (*c.* 500–428 BCE) who considered that the Moon was a stone. He thought that the Sun was a red-hot mass of iron, bigger than the Peloponnesus, the southern region of Greece that is about the size of Sicily. This idea that the Sun might be made of iron was based on a reasonable interpretation of the available evidence. An iron meteorite had fallen in about 467 BCE in ancient Thrace and Anaxagoras concluded that the visitor had come from the Sun. He was banished from Athens because his views about the composition of the Sun and the Moon were considered to be heretical. Little of his work has survived, but apparently he pictured the Earth at the center of a sort of large cosmic whirlpool. In this he anticipated the notions of Descartes in the sixteenth century, demonstrating the truism that few ideas are truly original.

The great Greek philosophers, Plato and Aristotle, whose philosophy has formed much of the basis for western culture, were mostly concerned with questions of purpose. Four elements, earth, air, fire and water, sufficed to make up the Earth. The heavenly bodies, in contrast, were composed of shining crystal, a perfect fifth element, or quintessence. The Moon was also made of this. The dark patches that one could easily see on the face of the Moon were thought to be the reflections in this perfect mirror from the mountains and oceans on the Earth. The doctrine of Socrates (*c.* 470–399 BCE) left no room for any changes or evolution and so did little to encourage scientific investigation. Plato (*c.* 428–347 BCE) concerned himself with the motions of the planets rather than their origin. In his scheme, the heavenly bodies were supposed to move in perfect circles.

The problem of perfectly circular orbits continued to haunt astronomers as late as Copernicus, over 1000 years later, until Kepler finally broke the spell. Aristotle (384–322 BCE) also thought that the heavens were permanent and thus not subject to the Earthly laws of physics as he perceived them. Both Aristotle and Plato did not believe in the plurality of worlds nor of life outside the Earth. The speculations of Aristotle were to dominate Western culture for two thousand years because they became theological dogma, something that was not the fault of the philosopher. The Ancient Greeks had discovered both inductive and deductive reasoning, Plato in particular favoring the latter approach, reasoning from first principles. These are, however, somewhat difficult to establish as any scientist soon discovers. Empirical observations dominate most planetary studies and uncomfortable facts destroy the most beautiful of theories.

Aristarchus of Samos, who lived around 250 BCE, proposed a refreshing contrast to the views of Aristotle. He placed the Sun at the center of the solar system, and included the Earth with the rest of the planets. He realized that the Earth was small in relation to the Sun. Not everyone today has made that intellectual leap. Aristarchus appears to be the first person to suggest that the Earth both rotates and revolves around the Sun. This idea was not forgotten, but lay around until being revived by Copernicus over a millennium later. It is fitting that a prominent crater on the Moon is named for Aristarchus.

Epicurus (*c.* 341–270 BCE), who was a strong critic of the views of Aristotle, did not give the heavens any special or separate status. He adopted the ideas of Democritus (*c.* 460–370 BCE) of a plurality of worlds and supposed that the heavenly bodies formed by random collisions of atoms, whose existence had been proposed by Democritus 150 years earlier. We would now call Epicurus a materialist.

The Epicurean School rejected divine explanations, and believed in physical causes. Unfortunately it did not encourage investigations into natural phenomena, so that no scientific advances resulted. Epicurean philosophy was mostly concerned with freedom and happiness and was very popular. It survived until the fourth century CE before the Christians managed to defeat it. Our best surviving statement of the physical theory of Epicurus comes from the Roman poet and philosopher Lucretius (*c.* 99–55 BCE). In his long poem *De rerum natura* (*On the Nature of Things*) he adopted many of the ideas of Epicurus. He encouraged a materialistic outlook and discouraged superstition. It is refreshing that he paid little attention to astrology, which was popular then as now. The poem was lost for 1000 years before being rediscovered by Poggius Florenitinus (Poggio of Florence) in 1417 CE. What path would the history of the world have taken if the ideas of Epicurus and Lucretius had taken root rather than those of the opposition?

Among others deserving a special mention, Eratosthenes (276–195? BCE) correctly calculated the radius of the Earth. His answer to this classical problem was within about 1% of the modern value, a technical feat that was not rivaled for the next 1500 years.

Ptolemy is famous for his theory of the solar system. He compiled a summary of Greek astronomical thought and data in his book, the *Almagest*. It was a triumph of the use of geometry in understanding the solar system. This book was the definitive work on astronomy until the end of the Middle Ages, and so remained the acceptable explanation for over a millennium. Like Lucretius, very little is known of his life, except that he lived in the second century CE. The works of Ptolemy were much studied by the later Arab astronomers. His birth and death dates are unknown, although Arab sources recorded that he lived for 78 years.

Nevertheless, in spite of his great reputation, Ptolemy remains an obscure figure. It is not clear how reliable his measurements were, particularly since he worked for the state religion, which was heavily concerned with astrology. He seems to have been endowed with bad judgment, since he rejected both the Sun-centered solar system of Aristarchus and the essentially correct value for the size of the Earth that Eratosthenes had worked out. Both decisions put the progress of scientific knowledge back for the next 1500 years. Perhaps Ptolemy's major achievement was to salvage the star catalogue of Hipparchus, the greatest of the ancient observational astronomers. He worked in the second century BCE and his catalogue listed 850 stars arranged in six orders of apparent brightness, more or less in line with modern concepts.

Like Plato and Aristotle, Ptolemy felt that the imperfect Earth could not be given a place among the heavenly bodies, which were composed of shining crystal in their cosmologies. Echoes of this philosophical approach still appear in the very common tendency to consider unknown or distant regions as uniform in composition. Examples include the deep interior of the Earth, the solar nebula and the universe, all of which were thought until quite modern times to be uniform; more recent information is rapidly dispelling these myths.

The system devised by Ptolemy placed the Earth at the center of the universe. The motions of the planets followed extremely complicated paths. In spite of its theoretical defects, it was a practical success. However, many of its problems had been long understood by skeptical observers. One of these was Alfonso X (The Wise) King of Castille (1221–1284 CE), who is commemorated by having one of the larger craters on the Moon named in his honor.

Laplace (1749–1827) in his *System of the World* tells the following story about him: "Alfonso was one of the first sovereigns who encouraged the revival of astronomy in Europe. This science can reckon but few such zealous protectors; but he was ill seconded by the astronomers whom he had assembled at a considerable expense and the tables which they published did not answer to the great cost they had occasioned. Endowed with a correct judgment, Alfonso was shocked at the confusion of the circles, in which the celestial bodies were supposed to move; he felt that the expedients employed by nature ought to be simpler. 'If the Deity' said he, 'had asked my advice, these things would have been better arranged'" [7].

In spite of such opinions, scientific knowledge in Europe by the fourteenth century was less advanced than in Greece and Alexandria in the second and third century BCE. The level of mathematics was at about that which the Babylonians had achieved two millennia before.

Following the collapse of the Roman Empire, there followed a thousand years of intellectual stagnation in the West. The Greek philosopher, Proclus is reported to have made "the last astronomical observation in the ancient world in AD 475." [5]. He is commemorated by the young crater, Proclus, on the Moon.

Astronomy survived through the work of Arab observers such as al-Farghānī; the crater Alfraganus on the Moon is named after him. Many of our brightest stars such as Aldebaran still bear their Arabic names. The great flourishing of Arabic science, particularly in mathematics, took place in the ninth, tenth and eleventh centuries CE. The revival of learning in Europe during the Renaissance led to the Copernican revolution in the sixteenth century. This created a new worldview that the human ego is still trying to come to grips with.

The Copernican revolution

The Copernican revolution is usually dated at 1543. This was the year of publication of the great work of Nicolaus Copernicus (1473–1543) *De revolutionibus orbium coelestium, libri VI (On the Revolutions of the Celestial Spheres)*. He is reputed to have received the book on the day he died. As a canon of the Church he had tenure but was apparently reluctant to publish. Few modern authors would care to wait so long.

The model of Ptolemy had placed the Earth at the center of the universe. This was obvious to everyone and equally agreeable to the ego of *Homo sapiens*. After all, it was clear to casual observers that the Earth was flat and that the Sun, Moon, planets and stars all revolved around it. Any child could understand this medieval view of the universe. One is reminded of the current debate over creationism and intelligent design, other examples of simplistic and erroneous views of the world.

Furthermore, the Ptolemaic system, for all its complexity, worked well enough for practical matters, including navigation: Columbus used it. Minor problems were accommodated by complicated adjustments until a complex array of epicycles and the like, to which Alfonso had objected, encrusted the whole scheme.

Copernicus, however, placed the Sun at the center. Why did he do this? One can do little more than speculate, 400 years later, but he seems to have viewed the Sun-centered system as more intellectually satisfying than the Earth-centered model of Ptolemy. Daniel Boorstin (1914–2004) records that "Copernicus possessed an extraordinarily playful mind and a bold imagination" [8] and that his model was driven by aesthetic rather than scientific reasons: "one places the lamp in the center of the room".

It is curious that Copernicus did not refer to ideas of Aristarchus of Samos, who had proposed a sun-centered system eighteen centuries earlier. But like most educated people of the time, he was aware of the ideas of Aristarchus, just as the idea of evolution was around long before Charles Darwin provided the data that confirmed it. Copernicus indeed refers to Aristarchus in a draft copy of his work which does not appear in the published version [9]. Along with Alfonso, other thinkers in the Middle Ages, of whom Nicolas of Cusa (1401–1464) and Regiomontanus (1436–1476) were examples, had laid the intellectual framework for dismantling the old system [10].

The new scheme of Copernicus was not without its problems, and in fact did not work as well as Ptolemy's for practical applications. The planets remained in circular orbits, so Copernicus still had to use even more epicycles than Ptolemy to account for their motions. According to this notion, planets, like a trick cyclist, rotated around in small circles, or epicycles, as they progressed in their circular orbits around the Earth.

Epicycles were an obvious solution to the problems of the apparent loops in the motions of the planets as seen from the Earth. This is seen most easily for Mars, which after moving slowly eastward though the constellations, reverses its normal path and travels westwards, before resuming its slow eastward course among the fixed stars. We now know that this curious reversal that we observe is due to the Earth, with its orbital period of 365 days, overtaking Mars which takes 687 days to go around the Sun.

It took a long time after the death of Copernicus for the idea that the Earth goes around the Sun to be commonly accepted. In our age, Darwinian evolution is likewise taking some time to become established as the accepted worldview. Giordano Bruno (1548–1600) usually enters the stage at this point, often regarded as a scientific martyr for his views about a plurality of worlds. It seems more likely that he was condemned because his theological views about God conflicted with the official doctrine of the Catholic Church. However, "far from being a martyr to science, Bruno actually harmed it, because the storm he raised caused the religious authorities to associate the Copernican system with anti-Christian agitation" [11].

The next significant step in understanding the solar system was taken by Tycho Brahe (1546–1601), another outstanding figure of Renaissance science. His chief accomplishment was the precise measurement of planetary positions. This was carried out by visual observation, as the telescope had not yet been invented. His observatory was on the island of Hven, a short sail from Copenhagen. Tycho was also concerned about problems with the complicated system of Ptolemy. So he produced a model in which the Sun and the Moon indeed went around the Earth, as everyone could see. However, he had the other planets rotate around the Sun. In this way, he had a foot in both camps. This compromise cosmology was popular, since it appealed to common sense observations and did not conflict with theological dogma. Variations survived until late in the seventeenth century, finally vanishing as the motions of the planets became well understood.

Tycho had other problems. He lost part of his nose in a duel and wore one made of tin, for cosmetic reasons. He also disgraced himself in the eyes of his aristocratic family by marrying a peasant's daughter. Finally, he was so unpopular with the other residents of his island that they demolished his observatory when he lost royal favor and he had to move with his data to Prague, in 1597.

Here chance plays its role. Just in time, in 1600, another refugee arrived in Prague. Johannes Kepler (1571–1630) had been banished from the pleasant town of Graz in Austria, a victim of Catholic persecution. He became Tycho's assistant and succeeded him as Imperial Mathematician when Tycho died, suddenly, in 1601. Kepler thus inherited, or perhaps just took ("usurped" was his word), the boxes that contained Tycho's monumental observations. These data formed the basis for Kepler's basic discoveries of the laws of planetary motion. His great contribution was to get rid of the notion, that had survived since Aristotle, that the planetary orbits were circular. He discovered that the orbits were elliptical and became an advocate of the Copernican System.

However, like many other scientists, he was mainly concerned with other matters so that, as one author has commented, the three major advances in Kepler's works on astronomy were concealed in a wilderness of nonsense. It is difficult to imagine the intellectual climate in which he lived. His mother was accused of witchcraft and he spent several years defending her, ultimately successfully, from the appalling fate that accompanied conviction.

In spite of such distractions and with a vast amount of labor, Kepler was able to fit the orbits of the planets into spheres based on the five "perfect" geometrical solids; cube, tetrahedron, octahedron, icosahedron and dodecahedron. They had long fascinated philosophers and the concept was widely known in Kepler's time. Plato had used the first four forms as the basic shapes for Earth, air, fire and water, while the dodecahedron was the model for the heavens.

Kepler considered that his calculations had answered a fundamental question: why were there only six planets (as known at the time), with five intervals between them? This cosmic limit was imposed because of the small number of "perfect" solid forms. However, the planetary orbits, on the basis of Kepler's own laws, turned out to be elliptical, not circular. Thus his elaborate geometrical system fell into ruins.

Clocks had been prominent features in town squares in Europe since the fourteenth century. Richard of Wallingford constructed one of the earliest, in about 1320, at St Albans Cathedral in England, during the reign of Edward III. They became more sophisticated as clockwork was perfected, and often included astronomical models as well as religious displays. A famous example is the great clock at Strasbourg, dating from about 1364. Others were at Mantua, Padua, Prague and Venice. Such mechanical marvels led to the idea that perhaps the universe was some kind of giant clockwork. A clock requires a builder, suggesting that a master craftsman had created the universe.

Once the solar system had been constructed by an omnipotent clockmaker and the system was set running, no further attention was needed. It would continue to operate under the laws of physics. Such ideas went back to Nicole Oresme (1325–1382), a bishop who had conceived of God as the master clockmaker. Kepler was an enthusiastic supporter, suggesting that perhaps magnetism was the driving force, just as falling weights drove Earthly clocks.

The clockwork idea was also consistent with the Bible. An Irishman, Archbishop Ussher (1581–1656) calculated that the creation of the world (including the universe) had occurred in 4004 BCE, on Sunday, October 23 at 9.00 am. Although Ussher gets the credit, the date and time are sometimes attributed to Dr John Lightfoot (1602–1675), Vice-Chancellor of the University of Cambridge, although this notion may be a later Victorian addition. This date, although now derided, was carefully derived from the available biblical record. What it represents is essentially that of recorded history. It was generally accepted at the time, even now appearing in many editions of the Bible. Shakespeare was aware of it, for in *As You Like it* (1599) Rosalind says, "The poor world is almost six thousand years old."

The significance of this date was that the universe had not had much time to evolve and everything must have been created in the beginning, more or less as it appeared now. Although more recent attempts by creationists to revive this date are usually dismissed as ludicrous, they should be strongly attacked as nonsense rather than being ignored.

The Copernican Revolution did not resemble those of more modern times. Fifty years after the publication of his system by Copernicus, little had changed. His ideas had disturbed neither the public nor the church. What was needed was some crucial observation to decide between Copernicus and Ptolemy.

This came, as is usual in scientific progress, with a technical advance. The telescope had been invented in around 1600 by Hans Lippershey (1570–1619), a Dutch spectacle-maker. Although he did not receive a patent, historical research credits him with this world-changing discovery. When the news reached Italy, the Senate of Venice asked Galilei Galileo (1564–1642), a skilled maker of instruments, to make some. He was the son of a lute player and composer, but had decided not to follow his father's career. We are still living with the consequences of that decision. It was not of course the intent of the Venetian state to upset the accepted view of the world. Their reasons were more down to Earth. Telescopes would obviously be useful for an empire based on sea power. One is reminded that the British Admiralty did not send out HMS Beagle, carrying Charles Darwin, because they wished to change our view of nature or overturn the authority of the scriptures. They needed better charts of the South American coast.

Galileo's observations are famous. The Milky Way was composed of stars and so maybe the universe was infinite. The Moon was not a smooth mirror after all, but rough and mountainous like the Earth and so perhaps made of the same material. Venus showed phases like the Moon, including a full face. This told Galileo that Venus was passing behind the Sun. Another critical observation that led to the collapse of the Ptolemaic System came when Galileo discovered, in 1610, that four satellites were rotating around Jupiter.

Copernicus and Aristarchus were right after all. The idea that the Sun, rather than the Earth, was at the center of the universe, caused a profound change in the view of our place in the world. It created the philosophical climate in which we live. It is not clear that everyone has yet come to grips with the idea, for we still cherish the idea that we are special and that the entire universe was designed for us.

Rene Descartes (1596–1650) then took up the challenge of the origin of the solar system. His view of the world was a completely mechanical one. He postulated that there was no basic difference between the forces driving a clock, the solar system or living matter. He proposed that the universe contained many circular eddies. Like a whirlpool, matter accumulated in the center of the vortex to form the Sun. Coarser particles were captured to form the planets. Satellites formed in secondary whirlpools surrounding the planets. He appears to have disregarded some of the conventions of his age, if John Aubrey's (1626–1697) biographical sketch is a realistic account. Amongst other gossip in Aubrey's *Brief Lives* is the statement that Descartes "was too wise a man to encumber himself with a wife, but he kept a handsome woman, by whom he had some children"[12].

By the time that Isaac Newton (1647–1727) appeared, the Copernican System had long dominated thought. Newton's work was the culmination of the work of Copernicus, Kepler and Galileo. Writing in 1704, Newton was impressed by the tidy nature of the solar system. He was irritated by the qualitative notions of Descartes, and showed that exact physical laws could deal with the complexity of the solar system. The planets were securely tucked into their orbits and the space between was apparently clean.

Newton assumed, at least publicly, that the world had been created essentially in its present form only a few thousand years before, according to the biblical timescale. This left no time for the system to evolve from a more primitive state, as Descartes had imagined. Thus it required a Creator, who had ordered each planet to move in its particular orbit.

The success of Newtonian mechanics reinforced the notion that the solar system was some type of celestial clockwork. This theme of a celestial clockmaker came to dominate thinking about the solar system in the seventeenth and eighteenth centuries. These ideas bore fruit in the construction of mechanical models of the solar system.

Attempts to model the solar system date back to antiquity. Cicero (106–43 BCE) in *De republica*, tells of having seen a model that Archimedes (287–212 BCE) had built. It showed the Sun, Moon and the five planets known to the Ancients. The Antikythera Mechanism, made in about 150 BCE and recovered from a shipwreck near Crete, also seems to be an ancient example of an orrery. They are named orreries after the Fourth Earl of Orrery, Charles Boyle (1676–1731), who had one of the first modern examples built.

These instruments became very popular. There is a beautiful example in the Meteorite Hall of the Natural History Museum in Vienna of a "Kopernikanische Planetenmaschin", made in 1761 for the Austrian Emperor. When Louis XV (1710–1774) constructed a new wing at Versailles, an orrery was placed in the central room, in contrast to the chapel which forms the center of the old wing. This was in keeping with the philosophy of the Age of Enlightenment.

However, Newton noted that there were small variations in planetary orbits, so in his system God had to intervene from time to time to make periodic repairs or adjustments, in effect winding up the clockwork. This led to complaints by his great rival, Leibnitz (1646–1716) that Newton was guilty of heresy, by supposing that God could have created something less than perfect. Certainly, given supreme powers, the construction of a well-ordered planetary system should not be beyond the powers of a competent clockmaker. Surely God would not have constructed an imperfect system and would have had enough foresight to create perpetual motion, rather than acting as a maintenance man who had to wind up the clock and make fine adjustments to the planetary orbits.

A little later, the great philosopher, Immanuel Kant (1724–1804) considered the problems of the solar system. In his time, philosophers worked on such important problems. He provided a correct explanation for the Milky Way, proposing that it was an edge-on view of a disk of stars. His suggestion that the fuzzy, lentil-shaped nebulae were distant island universes similar to the Milky Way showed remarkable foresight. This was a leap in understanding that was not confirmed until the third decade of the twentieth century, nearly 200 years later. These essentially correct insights perhaps explain why he gets so much credit for his ideas about the origin of the solar system.

Kant felt that the solar system could not arise purely from the mechanical ideas of Newton, but that God had to guide the design. Once this perfect plan was set up, the universe had no freedom to deviate from it. Kant's model for the origin of the solar system was based heavily on an analogy with the galaxies. It began with a chaotic distribution of particles. This material was assumed to be rotating and to develop into flattened rotating disks. The Sun formed at the center, and the planets formed at secondary condensations within the disk. He believed that "most planets are certainly inhabited" [10, p. 143]. Kant postulated the existence of many additional planets outside the orbit of Saturn, with a gradual transition to the comets. In his book, *General Natural History and Theory of the Heavens (Allgemeine Naturgeschichte und Theorie des Himmels)*, he populated all the planets with intelligent creatures. They became more clever with distance from the Sun. Thus a monkey on Saturn would be smarter than Newton, but an inhabitant of Mercury would be incredibly stupid.

When his ideas on the origin of the solar system are examined more critically, they turn out to be mostly vague statements. The many contradictions in Kant's hypothesis do not agree with the popular acclaim that it has received. Perhaps this is due to his eminence as a philosopher. It shows how difficult it is to account for the solar system, when one of the foremost thinkers of the Enlightenment failed to produce an acceptable explanation. His model is often linked, incorrectly, with that of Laplace, to whom we now turn.

LAPLACE AND HIS FOLLOWERS

We can date modern thinking about the origin of the solar system from the appearance, in 1796, of the System of the World by Pierre-Simon, marquis de Laplace (1749–1827) (Figure 2).

His work on celestial mechanics, although less well known in the English-speaking world, rivals that of Newton. Laplace was impressed, as Newton had been earlier, with the regularities in the solar system as it was known in the late eighteenth century. The planets all lay in a plane, and they all moved in the same anticlockwise direction around the Sun. The satellites revolved around their parent planets in the same direction. Laplace ignored the inconvenient fact that at least two satellites of Uranus, discovered by William Herschel (1738–1822) in 1787, were orbiting in a plane perpendicular to the rest of the solar system. Details often have other explanations; wisdom consists in knowing what is critical.

The orbits of the planets, although elliptical as every schoolchild is now told, are in fact nearly circular. This regular arrangement led Laplace to the concept that the system had arisen far in the past from a primitive rotating cloud, the "solar nebula". This idea has survived. This was in contrast to the ideas of Newton, who had apparently believed that the solar system had been created in its present form only a few thousand years earlier.

FIGURE 2 Laplace. Pierre Simon, marquis de Laplace (1749–1827).

Laplace, however, was an inhabitant of the Age of Enlightenment. Born into what we would now call a middle-class farming family, he had survived the French Revolution and was a distinguished member of the French scientific establishment at the beginning of the nineteenth century. He was able to show that the apparent variations in the orbits of the planets were self-correcting, and so God was not needed to adjust the system. Laplace gave a copy of his famous book to Napoleon, to whom he had taught mathematics when the Emperor had been an artillery cadet. Bonaparte, seeing no mention of God, presumably the designer of the system, asked Laplace about this omission. Laplace, having solved the problem that had bothered Newton, made his famous reply that he had "no need for that hypothesis"[13].

A watershed had been crossed. Now the solar system could be considered as having arisen by the operation of natural processes from a primitive beginning, rather than having been created perfect in the instant. This marks the beginning of modern attempts to understand how the Sun and the planets came into being.

At the same time that Laplace was writing his *System of the World* (1796), Josef Haydn (1732–1809) was composing *The Creation*, which he also finished in 1796. It was first performed in Vienna in April, 1798. This oratorio, for five soloists, choir and orchestra, takes over two hours to perform. It is the finest musical statement on the origin of the solar system and must be considered as one of the major achievements of western civilization. Haydn's sources were *Paradise Lost*, published in 1667 by John Milton (1608–1674), and the biblical account in Genesis that begins the Old Testament account in the King James Version of The Bible (1611) [14].

So Haydn had to form the Earth, its inhabitants both animal and human, and the heavenly firmament within the seven days allowed by the authors of the Book of Genesis. More recent work has relaxed this tight time frame, so that nearly 14 billion years are now available in which to reach our present position. We are still waiting for artistic statements of the stature of *The Creation* that incorporate our new understanding. This work among others is sometimes cited as evidence that religion is needed to inspire the creation of great works of art. This charge is of course true, but in its absence, artistic works would merely have taken a different course. One example must suffice. Haydn indeed wrote his magnificent oratorio, *The Creation*, based on the account given in Genesis. But this composer wrote an equally magnificent oratorio, *The Seasons* (1801), on a secular theme.

The egotistical and apparently obvious idea that the Earth is the center of the universe no longer attracts any scientific attention. This is not only because such notions have been replaced by those of the

Copernican Revolution, but because in such models, the origin of the Earth, Sun and planets was tied to the origin of the universe. After all, the Earth could hardly be younger than the rest if it occupied the central position. Now it has been established that the age of the solar system is only about one third of the age of the observable universe. This makes it no longer necessary, as was the case with the authors of the Book of Genesis, to seek an origin for Earth, Sun, Moon and stars in that order. Most of this progress has been made by the discovery of new facts, not by theories. Galileo's observations, like those of Darwin, have done more to give us a correct view of the world than most of the theorizing about it over the centuries.

2 The universe

"The true constitution of the universe–the most important and admirable problem that there is" [1].

AN EXPANDING VIEW

In order to obtain some perspective on the place of planets in the scheme of things, it is useful to contemplate the scale of the universe, as we perceive it at present. Everything in the universe is very isolated. The nearest star to us is Proxima Centauri, a red dwarf and the faintest member of a triple star system of which Alpha Centauri is the brightest. This star is familiar to dwellers in the southern hemisphere, as it forms one of the Pointers to the Southern Cross. Light from this nearest star takes 4.2 years to reach us. Although Proxima Centauri is the nearest star at present, the dwarf star Ross 248 will succeed to the title in about 33,000 Earth years. Because of the slow relative movements of the stars, our familiar constellations, such as Orion the Hunter and his companion, the Great Dog, will be rearranged and replaced by other groupings in the future. Edmund Halley (1656–1742), of comet fame, seems to have been one of the first to realize this, by observing that the positions of many stars in the early eighteenth century differed from those recorded in the catalogue of Hipparchus in the second century BCE.

The Milky Way, a spiral galaxy with a bar-shaped core, is about 100,000 light years in diameter and is rotating slowly. The solar system, and we, reside in one of the dusty and gas-rich spiral arms (the Orion-Cygnus Arm) which is about 28,000 light years from the center. The galaxy turns in a slow, majestic wheeling motion. It has rotated less than 20 times since the solar system formed, as it takes about 250 million years to make one revolution. The grand scale of this movement can only be appreciated on geological timescales. Two hundred and fifty million years ago, the Paleozoic Era was drawing to a close, a time marked by one of the great extinctions at the boundary between the Permian and the Triassic (see Appendices), when over 95% of life on Earth was extinguished, including the trilobites that had flourished for 300 million years.

The nearest major galaxy, M31 or Andromeda, is 2.5 million light years distant, and forms one of at least 30 members of the Local Group of galaxies. These include the Magellanic Clouds that are clearly visible in the southern hemisphere. They were seen by and are named for the Portuguese navigator, Ferdinand Magellan (1480?–1521). He commanded, but unfortunately did not survive, the first expedition to circumnavigate the Earth. Beyond the Local Group, the universe extends as an apparently endless array of over 170 billion galaxies.

The expansion of the universe

The discovery in 1929 by Edwin Hubble (1889–1953) that the universe was expanding is now known to all. But there was speculation about whether the universe would keep expanding, reach a steady state or collapse back on itself. Now we know that not only is it expanding, but also to much surprise that the rate of expansion is accelerating. This result seems established firmly enough to satisfy the Nobel Prize committees. It is thought to be due to an unknown form of energy, a so-called "dark energy" that overcomes gravity and accounts for about 72% of the mass-energy of the universe. Interesting times indeed. However, this is the business of cosmologists, not a student of rocks or planets.

What about the remaining 28%? The values are provocative. The wonderful array of elements in the Periodic Table amounts to a bit less than 5%. Most of it is hydrogen dispersed through space. Stars, that at least we can see, make up less than 0.5%. Planets contribute only a trivial amount, although it is important to us since we are standing on some of it. But the effect of gravity on what we can see demands the presence of undetectable (at present) "dark matter" which amounts to about 23% of the universe. It is notoriously elusive.

Many candidates have been put forward. One suggestion was that it was the burnt-out remains of massive stars formed at an early stage. Other possibilities have been very faint stars, brown dwarfs and such creatures, but there is a scarcity of all of these. These are lumped together as "massive astronomical compact halo objects" or MACHOs, another example of the acronyms that have come to plague us, as an apparently enduring legacy of the Second World War.

But none of this amounts to much and most dark matter is not formed of atoms, but may consist of "WIMPS" (Weakly Interacting Massive Particles), besides which leprechauns might seem to be firmly rooted in reality! However, we have been surprised before. The Earth is round, not flat and revolves around the Sun. Both notions appeared to be ludicrous propositions to primitive societies and even now have some supporters.

The "age of the universe"

The reason for raising the question of the age of the universe in a book on planets is to clarify the difference between the two. Although it now seems clear to everyone that planets are younger than the universe, this was not, however, always obvious to *Homo sapiens*. Few creation stories or religious explanations make any distinction. Thus in the story told in the Book of Genesis, the Earth appears first, followed later by the Sun, Moon and lastly by stars.

Although this was perhaps a satisfactory explanation for desert tribes, it is different from the order that science shows. The solar system formed late in the history of the universe and its origin is clearly the result of a separate event. It turns out that planets form readily, so that the formation of our solar system was just another commonplace event, only a little unusual in the array and near-circular orbits of our planets. The fundamental philosophical conclusion is that our system of planets arose by normal physical processes, as Laplace had perceived. Galaxies had been forming and stars had lived and died for several generations before the solar system appeared in a spiral arm of the Milky Way.

The Hubble "constant" measures the rate of recession of the galaxies and so gives the time from the "origin" of the universe in the Big Bang. After 70 years of intense effort, with values ranging between about 40 and 90, a value (within small errors) has been arrived at of about 72 kilometers per second per megaparsec (a megaparsec is 3.26 million light years). The "age of the universe", that is, the time since the Big Bang, has, after much controversy, thus been pinned down to 13.75 billion years. At any popular lecture on the solar system, one can guarantee that there will be two questions, the first about the Big Bang and the second about UFOs. Here I defer to the excellent account of the Big Bang that is to be found in *The First Three Minutes* (1977) by Steven Weinberg (b. 1933).

Three pieces of evidence are generally used to support the presently accepted Big Bang model. One is a faint glow, the embers from the high temperatures of the Big Bang, that is diffused through space. Because of the expansion, this apparent remnant from the Big Bang is now observed red-shifted to wavelengths at around one centimeter in the microwave region. The temperature has dropped to 2.725 Kelvin, close to absolute zero. The second piece of evidence in favor of the Big Bang is that the abundances of deuterium, helium and lithium that we observe have been held to agree with the predictions of that theory. The third piece of evidence is the entertaining observation that the sky is dark at night.

The darkness of the night sky

Why is the sky dark at night? Corin, the shepherd in Shakespeare's *As You Like It* (1599), knew that "a great cause of the night is lack of the Sun" [2], but the question is a little more complex. If the universe is infinite and filled with stars, then every line of sight must eventually intercept a star. Accordingly the night sky, and the daylight sky for that matter, should be ablaze with stars.

The problem had been around for a long time. Thomas Digges (1546?–1595), writing in England in 1576, raised the question. He believed that the universe was infinite and that absorption of light from distant stars was responsible for the dark night sky. Kepler thought about the problem, deciding that it showed that the universe could not be infinite. What we were looking at between the stars was the darkness outside the universe.

But the problem only became famous when the German astronomer, Heinrich Olbers (1758–1840), publicized it in 1823, so it became "Olbers' Paradox". We are fortunate that Edward Harrison (1919–2007) has written an elegant synthesis, *Darkness at Night, a Riddle of the Universe* (1987), which lists the many ingenious solutions proposed for this problem.

There are several reasons why the night sky is dark. Stars have restricted lifetimes and burn out over periods ranging from millions of years for massive ones to billions for dwarfs. The expansion of the universe since the Big Bang has spread out the galaxies and stars. Light from more distant stars and galaxies has been shifted to longer wavelengths outside the visible range. Since the universe that we see is nearly 14 billion years old, light from more distant regions has not had time to reach us. At the beginning, the sky would have been brilliant.

Early insights into the correct solution were proposed by Lord Kelvin and curiously enough by the author, Edgar Allan Poe (1809–1849) in his poem *Eureka*. Clearly one should listen to the poets.

However, it would be premature to suppose that we have an ultimate solution to the fundamental problem of the origin of the universe. The notion that the universe began at some definable point has always been philosophically unsatisfactory. The Big Bang has been called an event without a cause. However, despite such problems, it represents the only currently acceptable scientific explanation for the origin of the universe. The question is open, like much else in cosmology. Now it is high time for this student of the planets to return to a more familiar neighborhood.

Galaxies

These are among the major components of the visual universe and need to be mentioned here, as the planetary systems reside within them. Like many objects, they would not have been predicted if they hadn't been observed. How these beautiful structures arose from the primordial soup of fundamental particles is one of the major quests of cosmology. In the inflationary scenarios they perhaps began from minor fluctuations as the Big Bang fireball expanded.

Typically they contain 100 billion stars, although like most things in nature they display an infinite variety. Most of the nearby examples at least fall into only a few general categories (ellipticals, spirals, barred spirals and dwarfs) as they progressively lose their bulges and transform into more aesthetically pleasing disks. Most contain a black hole at their center, the final triumph of gravity. Elliptical galaxies tend to have older stars than spirals. Former models for the formation of a spiral galaxy, including our own, began with a spherical mass of gas in which stars formed. The sphere collapsed to a rotating disk within a few hundred million years, leaving a halo of globular clusters of stars to outline its original extent. In this model, the evolution of the Milky Way galaxy began with the formation of the halo at around 12 billion years ago. As the galaxy collapsed to a disk, star formation began in the spiral arms around 10 billion years ago. Although it was previously thought that most galaxies formed very early in the history of the universe, it is now clear that galaxy formation has been a continuing process. So the splendid galaxies that we admire are not permanent, but evolve with time.

It now appears that galaxies have a much more complex history than previously imagined. Like the terrestrial continents, they seem to be composed of many separate units that have been swept together.

Thus, galaxies probably do not evolve in isolation, but have undergone many collisions.

Galaxies lie along chains, sheets, filaments and knots in patterns that perhaps represent spacings from an earlier pre-galactic time. The largest sheet-like structure that has been observed is the "Great Wall," which contains many thousands of galaxies and is over 500 million light years long.

In many cases, galaxies appear to be located on the surfaces of spherical shells surrounding dark regions apparently devoid of galaxies. Such structures have been compared to soap bubbles. Two-dimensional pictures of galactic distribution look like filaments, apparently because the galaxies congregate around the edges of the very large empty bubbles. These great bubbles are about 150 million light years across and are apparently empty. Clusters of galaxies appear to have formed within the last few billion years and these clusters are now beginning to be bound by gravity into superclusters. The universe seems to be forming itself into larger and larger units as it gets older, the same trend that empires on Earth have followed, to their ultimate ruin.

Not all galaxies are suitable in which to form Earth-like planets. Our small dwarf satellite galaxies, the Magellanic Clouds, are poor in metals, perhaps because they are old. The position within a galaxy also matters. Near the center, radiation levels are high. The outer parts of our galaxy are mostly poor in metals, stars having less than 10% of the metal content of our Sun.

Our solar system seems to be situated in a favored location and rotates about the galaxy in a nearly circular orbit. Indeed the Milky Way galaxy seems relatively quiet with a low rate of star formation, perhaps a good place in which to find ourselves.

Giant molecular clouds

Dark clouds of gas and dust of various sizes occur commonly in the spiral arms of our galaxy. Of most interest here are the "giant molecular clouds" (so called because most of the hydrogen is present as the molecule H_2). They are the most massive objects in the galaxy containing about half the mass of gas and dust, and are between 100 and 1000 times denser than the interstellar medium which contains about one hydrogen atom per cubic centimeter. Yet they occupy only a fraction of space in a galaxy, 1% or so. They appear to form by turbulence from the interstellar medium and their ragged structures are as chaotic as storm clouds, complete with filaments, sheets and clumps. They have brief lives, perhaps not much more than 10 million years and they inherit much instability from the interstellar medium.

They may be up to 100 light years in diameter and contain enough gas to form between a hundred thousand and several million stars, but are cold, only about 10 K. The classic example of a giant molecular cloud is the Orion nebula that forms the middle "star" of the sword of Orion, the Hunter. In these giant clouds, nature has managed to form many complex molecules, mostly compounds of hydrogen, oxygen, nitrogen and carbon. Over 150 different organic compounds have been identified. There is enough ethyl alcohol, the common intoxicant, out there, not only to sink a ship, but also to drown the Earth.

The clouds have a hierarchical structure, containing many smaller clumps, typically a tenth of a light year across and at about the mass of the Sun. These rotating clumps are the sites of star formation as they collapse under gravity. But the radiation from stars, particularly from the massive O type, in turn destroys their parents, so that the gas in the molecular clouds eventually returns to space.

How to build a star

The conversion of gas into stars is a process that has long been amenable to physics and so has led to the remarkable expansion of astrophysics. Two steps, both about a century old, but that have stood the test of time, seem especially important. The first was the classification of stars into OBAFGK and M types, easily remembered by the famous mnemonic "oh be a fine girl, kiss me" but spoiled more recently by the addition of L, T and Y classes to accommodate the long sought but newly discovered faint brown dwarfs [3].

The second, more significant step was the famous Hertzsprung–Russell diagram (Figure 1) which relates the brightness of a star to its surface temperature. The diagram has survived, like the Periodic Table of the Chemical Elements, because both are underpinned by basic physics, a fact of which their founders were not aware.

The formation of normal stars is one of the classical problems in astronomy. Although the basic physics is well enough understood, there still remain unsolved problems dealing with the birth of stars. Stars form at very different rates in different types of galaxies, depending on the amount of gas available. Our galaxy currently makes about four stars a year, but the rate varies widely among galaxies, depending on the supply of gas. Many elliptical galaxies have used up all the available gas and star formation has ceased. Others, particularly where two galaxies have collided and a fresh supply of gas has been acquired, produce stars in a sort of frenzy, appropriately labeled "star bursts". However, the basic process of star formation seems much the same. It is the rate that varies.

It takes about a hundred thousand years for the gas in a core to collapse due to gravity to the point where the pressure and temperature are high enough to turn on the nuclear furnace. Various classes of protostars, labeled from 0 to III, have attempted to define this evolutionary process, but the classification runs into the difficulty of observing such faintly glowing objects. As the fragment of the molecular cloud separates, it inherits some of the cloud's angular momentum. This prevents material falling directly into the forming star in the center. Instead the gas, ices and rock spin out into a disk. To conserve angular momentum, any mass falling into the star requires an outward transfer of angular momentum. It is for this reason that our planets contain 99% of the angular momentum of the solar system.

As the star grows, the disk spins up, reaching the point where material can no longer fall into the star. Is this why stars do not get infinitely big? Such simple questions, like other similar matters (why is the sky dark at night?), often conceal fundamental truths. But it is also a contest between gravity and heat. The ignition of the nuclear furnace occurs when the mass reaches about one third of the mass of the Sun. Strong outflows of gas and jets of material then break out from the turbulent young star. This also limits the infall of gas, while the heat generated supports the star against the crushing force of gravity. The sizes of stars are thus limited and they do not grow to an infinitely large mass. As the core collapses into the central star, the remaining gas and dust now forms a spiral disk that contains most of the angular momentum. Eventually, a central star is left surrounded by a rotating disk from which planets might form.

We see evidence of these disks and early strong winds in the T Tauri and FU Orionis stars. These young stars are less than a million years old. They are important because they tell us what the early Sun was like and so cover a crucial period in the development of the solar system. Many are surrounded by dusty disks [4].

Have stars always formed at a constant rate? Apparently not. From the fossil evidence in galaxies, it seems that star formation rates increased following the Big Bang, reaching a peak at around 3 billion years ago, and have been declining ever since. The universe is thus not a fixed and immutable entity but has been evolving, like much else.

Stars vary widely in mass from the lowly brown dwarfs, that extend into the planetary region, but which begin by convention at 12 or 13 times the mass of Jupiter, where temperatures and pressures are enough to initiate deuterium fusion. More normal stars begin with hydrogen to helium burning at 75–80 times the mass of Jupiter, forming the very common red dwarfs that constitute over 75% of all stars. Most stars form in clusters, although they soon separate so that few remain together after a million years.

The masses of stars range up to giants and supergiants, that can sometimes reach over 100 times the mass of our Sun. While the dimly glowing brown dwarfs can be difficult to spot, the supergiants can be hundreds of thousands of times brighter than our Sun. The number of stars falls off with increasing mass in a quite regular, perhaps universal, but not well-understood manner, called the Initial Mass Function (IMF). Small stars are much more common.

Brown dwarfs

Between the smallest red dwarf stars and large planets such as Jupiter lies the realm of the brown dwarfs. They need to be discussed here, as they have often been confused with planets and indeed overlap with them at the low mass end. A few examples are known where they have planets in orbit.

The canonical classification places them between 12–13 and 75–80 Jupiter masses. They are often thought to be intermediate between stars and planets, but this is a common misconception. Brown dwarfs are true, if they are minute, stars formed from the gravitational collapse of gas clouds but too small for the fusion of hydrogen to helium to occur; the defining characteristic of conventional stars. These dwarfs are usually found either wandering alone or as binary pairs, probably left-over remnants of the star-forming process.

Brown dwarfs occupy the lower right-hand corner of the Hertzprung–Russell diagram, a space in which there used to be a great temptation to place Jupiter. Being so small, they barely radiate and are as difficult to find as pygmies in a forest. The coolest so far found have surface temperatures of only 300 K, only a little warmer than Jupiter at 150 K, whose energy is mostly a residue from its accretion.

Good hunting grounds for brown dwarfs are to be found in the young star clusters. One is the Hyades cluster, a beautiful group in the constellation of Taurus, the Bull that lies between the Pleiades or Seven Sisters and the magnificent constellation of Orion, the Hunter. The stars in the Hyades formed at about 600 million years ago, just as life on Earth was undergoing an explosive diversity of forms and the first hard-shelled animals were making their appearance. These we see as trilobites and other species preserved as fossils in strata of Cambrian age, as well as the marvellously preserved soft-bodied forms of many extinct animals in the Burgess Shale, which was then mud on the ocean floor, but would eventually become part of the Rocky Mountains of British Columbia.

The splendid group of the Pleiades, about 400 light years away, is another good prospect for these elusive fellows. The stars in the Pleiades formed about 100 million years ago, when the dinosaurs were enjoying their heyday in the warm Cretaceous sunshine. It was also around this time when a large asteroid or comet hit the Moon, forming the great crater named after Tycho. It is noteworthy for its spectacular set of rays of dust flung out by the explosion, that go right across the face of the Moon and are easily visible in a small telescope or binoculars.

There are few bodies with more than about 12–13 Jupiter masses in orbit around stars. This lack became famous as the "Brown Dwarf Desert". However, to confound classifiers, a few massive objects have been discovered, existing in an oasis-like fashion in the desert. Corot 3b is a good example of the problem. Over 20 times more massive than Jupiter, it is in a 4-day orbit around a massive star. As it is far above the 13 Jupiter mass limit, is it a brown dwarf or a planet? A more extreme example is Corot 15b, 63 times the mass of Jupiter, with an orbital period of 3 days. To make the problem more complex, free-floating brown dwarfs with masses as low as three times the mass of Jupiter are known, through the work of the Spanish astronomer, Maria Rosa Zapatero Osorio amongst others. Are these small brown dwarfs or planets? Then there are the reports of huge numbers of "free-floating planets". So the possibility arises of overlaps between objects of any size, free-floating or in orbit, a true classifier's nightmare. So probably planets will end up by being defined as forming in orbit around a star and brown dwarfs as forming by condensation, like stars. So the upper mass limit for planets is a movable feast.

Red dwarfs

The smallest stars that undergo nuclear burning of hydrogen to helium are the red dwarfs, that range down to about one tenth of the mass of the Sun. They need to be discussed here as they form the vast majority of all stars, although most are too dim to be observed by eye. Even the largest have little more than 10% of the luminosity of the Sun. Their habitable zones are very close to the star. Apart from other difficulties, the planets in such locations would get locked into synchronous orbits, with one side too hot and the other too cold. Other problems are that red dwarfs frequently produce stellar flares, are very variable in luminosity and are powerful X-ray and ultraviolet light sources. Not a good environment in which life might flourish. Red dwarfs burn up their hydrogen fuel very slowly, so they have very long lives that may last up to trillions of years. Curiously they do not seem to be represented among the most ancient stars. All that have been found so far contain some metals. Either they formed after some element synthesis had occurred, or they have been polluted later by addition of metal-rich gas.

Binary stars

The cores from which stars form may inherit so much rotation that they may fragment. Thus instead of a single star, binary or more rarely, multiple star systems may form instead. They need to appear here as about 60% of stars have a companion, usually on a highly eccentric orbit. Indeed the distribution of eccentricities of exoplanets and binary stars is similar. This raises the question whether such eccentric orbits are a natural outcome in nature, so that the closely circular orbits in our planetary system become an outlier.

Although gravitational effects may disrupt the formation of planets around binary stars, planets are found orbiting one or both members. Stable habitable zones for planets may exist around either member of a widely spaced binary (separated by more than 50 AU) or encompassing both stars if they are close together (within 3 AU).

Metallicity

Although the composition of stars is dominated by hydrogen and helium, most contain "metals" that may reach concentrations up to 4%. One classification of stars distinguishes three classes, in order of decreasing metal content. They are conveniently labeled as Populations I, II or III.

Our Sun (with 1.4% metals) is a member of Population I which resides in the spiral arms of the galaxy. It is around these metal-rich stars that planets prefer to form. Although it has been suggested that these metal-rich stars are only apparently so and have been contaminated by metal-rich planets falling into them, it is easily calculated that this is a trivial effect. The metal content of a star, with or without planets, is an inherent property.

Population II stars, which have lower metal contents than population I, reside in the outer reaches of the galaxy and form much of the so-called "halo" population. The most extreme examples have over 100,000 times less metal than the Sun, but these are as difficult to find as needles in a haystack. Planets are scarcer around Population II stars. A search for transiting planets around 47,000 stars in one member of the galactic halo, the globular cluster 47 Tucanae, found nothing, although more than a dozen planets were expected.

The hypothetical Population III represent the earliest stars, which contain only H, He and a little lithium left over from the Big Bang. None have yet been found. They were likely very massive with short lives. Any star that has survived from that remote epoch is likely to have been polluted by some of the younger metal-rich gas, increasing the difficulty of discovering them.

OUR SUN: A COMMON STAR?

The Sun, a G class star lying on the Main Sequence of the Hertzsprung–Russell diagram, is a yellow dwarf that contains 99.9% of the mass of our solar system. Curiously enough, the fact that the Sun and other stars are composed mostly of gas in the form of hydrogen and helium was established less than 100 years ago. It is interesting that such a fundamental fact about the universe is such a recent discovery. Prior to that, iron was thought to be a dominant constituent, on account of the abundance of iron lines in the solar spectrum. This was not so far from what Anaxagoras had thought 24 centuries earlier.

How common are stars like the Sun? Most stars are smaller, the commonest being red dwarf M type stars that form over 75% of the stars in our galaxy. Ninety percent of nearby stars are less massive than the Sun. Although ages of stars are difficult to pin down with much precision, we know from the unique evidence in meteorites that the Sun is only slightly older than 4,567 million years. Fortunately it has a life of around 10 billion years, something that gave adequate time for intelligent life to develop. Unlike many stars, it is apparently quiet, allowing for stable planetary conditions. The Sun is also a single star, although most similar stars are members of binary systems.

The most recent estimate for the composition of the Sun [5] comprises about 1.4% of elements heavier than helium, commonly expressed as "metals" or Z, where X and Y are the abundances of hydrogen and helium respectively. The revision from the commonly accepted 2% is due to a downward revision in the abundances of C, N, O and Ne. The composition of the Sun appears a little unusual, with more metals than nearby stars so, ironically, the Sun does not appear to be a typical "solar-type" star.

In about another 5 billion years, the Sun will eventually come to the end of its life. Current models suggest that as the hydrogen in the core is used up, gravity will start to take over as the furnace shuts down. As the Sun collapses, the pressure inside builds up to the point where the furnace reignites, burning hydrogen in a shell outside the core. In the course of this, the Sun swells up to become a red giant, expanding out to the orbit of Venus within a few million years. However, it will have shed perhaps a quarter of its mass in the process. The Sun shrinks in volume again as the fire goes out and gravity takes over. As the temperature in the core reaches 100 million degrees, a second cycle of nuclear fusion, involving helium, begins. The Sun balloons out in a second red giant stage. During this time, such useful elements as carbon and oxygen are produced in the fiery furnace.

Perhaps the Earth will be swallowed at this time, although Jupiter and the outer planets may survive these events. Further catastrophes and flashes follow on the Sun, during which it sheds most of its mass. The elements it has so usefully produced are scattered into space, providing material for new stars. As the furnace finally shuts down, gravity will exert its overwhelming force and reduce the Sun to a geriatric white dwarf, about the size of the Earth. This tiny creature will be so dense that one cubic centimeter will weigh many tons, a challenge beyond the imagination of our weight lifters. The Sun is not massive enough to collapse into a black hole. That ultimate fate is reserved for much more massive stars.

The universe will continue on its way, heedless of the disturbance of a dying star and of the one-time existence of our planets. The elements forming the Sun and the planets will be dispersed into the interstellar medium before being incorporated into new stars and planets. Meanwhile, other events such as the predicted collision of our galaxy with the Andromeda Galaxy, which is now approaching us at 120 km per second and is due here in only 2 billion years, may upset the predictions of this model.

Laplace and his solar nebula

In the eighteenth century, scientists were unaware of the backward revolution of Venus, of the existence of minor satellites with exotic orbits, of the strange orbit of Pluto and its companions and of other irregularities. The solar system that they saw appeared as well ordered as a clock. The planets and satellites rotated in the same sense, both around the Sun and about their axes of rotation. They lay close to the plane in which the Earth rotates around the Sun, although the two satellites of Uranus known at the time rotated around their planet at right angles. The French astronomer and mathematician, Pierre Simon, marquis de Laplace, ignored that minor detail and proposed in 1796 that the solar system originated from a rotating disk of dust and gas. He called this disk the solar nebula. In his model, the planets condensed successively from rings as the nebula contracted.

This concept survived in its original form until late into the nineteenth century. The view that the Sun and the planets formed from a rotating disk of gas and dust, the solar nebula, now provides such an obvious explanation that it has become an ingrained truth. Laplace would no doubt have been pleased that his concept has survived and is now widely applied to disks around newly forming stars.

Increasing observations have enabled classifiers to separate disks into three classes: primordial, transitional and debris disks, although like many such attempts, it suffers from trying to divide a continuum. The first two gas-rich classes are what is generally understood as protoplanetary disks (sometimes called proplyds), while the gas-poor debris disks are typically found around mature stars and are analogs for our asteroid and Kuiper belts. Most of the dust in them, that creates a nuisance for planet seekers, is due to collisions and evaporation of the assorted debris [6]. Laplace had imagined that the planets condensed from rings in the nebula and such models formed the basis for more recent "Neolaplacian" models. But such debates have become much less relevant as it has become clear that planets do not form in situ, but migrate over considerable distances.

Gases, ices and dust

Aristotle and Plato needed five components to build the heavenly bodies. We have made some progress, needing only three. The great mixture of ingredients that form the material from which the stars and planets were built can be reduced to gas, ice and dust (loosely rock). The "gas" component was mostly hydrogen and helium, the two elements that form the bulk of the universe and constitute most of the mass of the initial solar nebula.

The remaining 1.4% in our disk consisted of various ices and dust composed of the heavier elements, the notorious "metals". These had slowly accumulated in the interstellar medium from nearly 10 billion years of element formation in previous generations of stars. The "ices" were mostly compounds of the abundant elements carbon, oxygen and nitrogen with hydrogen and were present at the low temperatures of the nebula as water ice, ammonia, carbon monoxide and methane ices.

"Dust" or "rock" was mostly a mixture of silicate minerals, sulfides and metal. The metal is mostly iron alloyed with some nickel, cobalt and traces of other very metallic elements such as platinum and iridium. The amount of rock in the nebula was so small that it could be ignored to a first approximation, except that we are standing on some of it. A corollary is that rocky planets are likely in general to be small; there is unlikely to be enough material in disks to make a giant rocky planet the size of Jupiter.

Our planets themselves likewise fall into three classes. The gas giants, Jupiter and Saturn, are mostly gas. We have found many gas giants, analogs of Jupiter, among the exoplanets. The ice giants, Uranus and Neptune, are mixtures of ice and rock with only a little gas, and analogs of these are present as well. They turn out to be surprisingly abundant. Many planets intermediate in size between the Earth and Neptune, labeled as super Earths, are also common among the exoplanets. Our familiar terrestrial planets are mostly rock with cores of metal, and such small rocky planets are very common among the exoplanets. How did we finish up with all this marvellous complexity from such unpromising material? How all this came to pass is one of the themes of this book.

It may seem surprising that we are well informed about the composition of our primitive solar nebula, which vanished so long ago into the Sun and planets. The composition of the Earth is of little use in this enquiry as the planet has had a complicated geological history. The rocks that we now find on the surface are the end product of over 4 billion years of geological processing and now contain only a faint and debatable record of their primitive beginnings.

Fortunately, some samples from these remote times have been delivered to us. Meteorites frequently fall on the Earth, displaced from the asteroid belt by tidal interactions, mostly with Jupiter. Among this debris from space, we are fortunate enough to have some meteorites, the famous carbonaceous chondrites, usually called CI, from Ivuna,

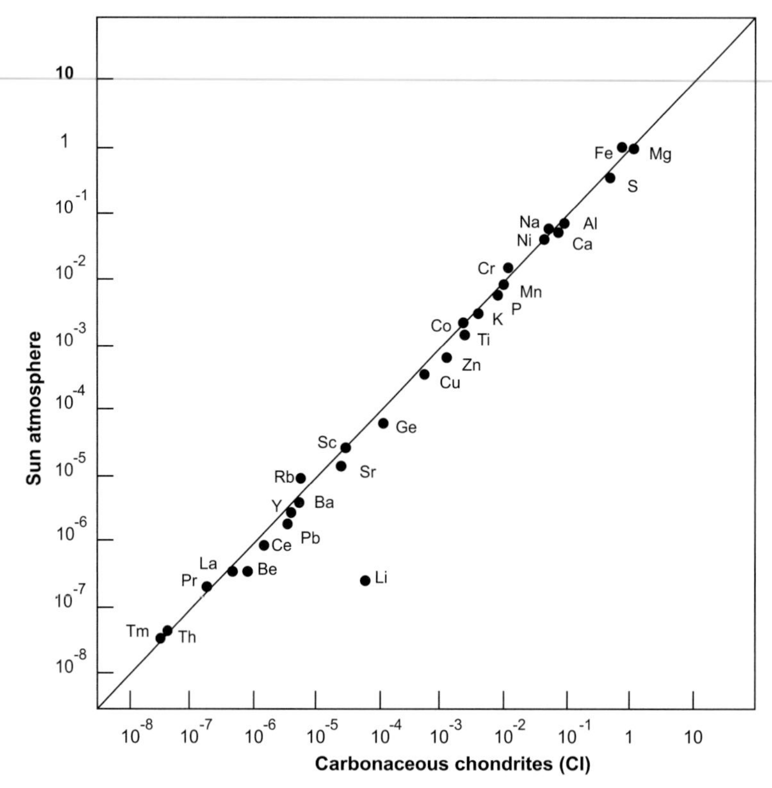

FIGURE 3 The close correlation between the composition of the "rocky" fraction of the Sun and the CI chondrites. This comparison extends to all elements except those present in the Sun as gases (H, He, rare gases) and ices (C,N, P and O). Lithium is depleted in the Sun, relative to CI, as it is consumed in nuclear reactions.

although the meteorite named Orgueil is the most celebrated member of this rare class. Their chemical composition is unchanged from the earliest times and matches closely the dust component of the original solar nebula (Figure 3).

How do we know this interesting fact? The rationale for this statement is that elements like sodium, potassium, lead, iron, magnesium and uranium in this class of meteorites, match the composition of these elements in the solar photosphere. The match is excellent

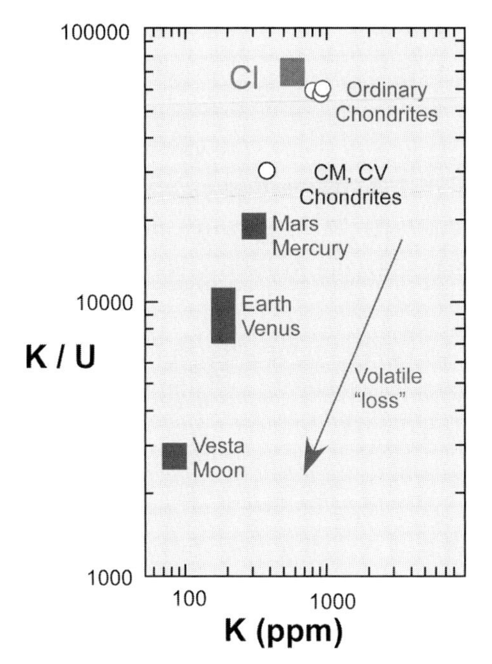

FIGURE 4 Volatile element depletion

The widespread depletion of volatile elements in the inner solar system, displayed here as the abundancc of potassium (K), a "volatile" element relative to uranium (U), a "refractory" element. CI gives the composition of the "rocky" fraction of the Sun and of the primitive solar nebula, as explained in the text. The chondrites are various classes of stony meteorites (courtesy Scott McLennan). See also color plates section.

for all elements that made up the dust component of the original nebula, except for lithium which is consumed in the nuclear furnace in the Sun. As the Sun contains 99.9% of the mass of the system, the CI composition is taken to reflect that of the "rock" fraction of the original solar nebula.

One might have thought that our Earth would be a prime example of this original dust or rock component of the nebula. It turns out, however, that the inner rocky planets, Earth, Venus and Mars, as well as most meteorites, have lost not only gases and ices, but also "volatile" elements such as sodium, potassium and lead. The entire inner nebula out to 3 or 4 AU was affected (Figure 4). The CI meteorites indeed lost gases and only retained a little water ice, but the other rocky elements were not affected. For this reason, they have retained the composition of the dust fraction of the nebula so that they match that of the Sun for the elements in the original dust.

Volatile and refractory elements

Everyone is familiar with chemical reactions on the surface of the Earth. Mainly it's a matter of forming compounds such as sodium chloride which every schoolchild knows about. The grouping of elements in that great triumph of nineteenth century science, the Periodic Table of the Chemical Elements, explains why, for example, sodium atoms join with those of chlorine to form table salt. It's a matter of how the outer shells of electrons of the atoms are arranged. But these properties are less important at temperatures hundreds of degrees hotter than the laboratory Bunsen burner, and pressures lower than the best vacuum that we can obtain.

Out in the universe, other properties take over. The most important is whether the elements have high or low melting and boiling points. Those that melt and boil at high temperatures are called refractory. For the geochemist or cosmochemist studying the cosmos, elements that boil and evaporate above 1000 to 1200 K are refractory. Examples are titanium, calcium or uranium. At the lower end lies the region of the volatile elements, that condense below 1000 K, including such familiar examples as lead, potassium, sodium, sulfur, copper and zinc. Although they are all nicely stable in our chemical laboratory, when temperatures get high enough, these elements are present as gas, leaving only the refractory elements condensed in solid form in grains. These matters are of little importance to astrophysicists studying stars, but are crucial to students of rocky planets.

Disks around stars: brief lives

This section is overly biased towards our own solar nebula which remains, for the present, our best example of a disk. Even in this well studied example, some detective work is needed to reconstruct events that occurred over 4.5 billion years ago. Our attempts to reconstruct what the nebula was like originally suffer from the same problems that historians have. One has to avoid folklore, over-interpretation and wishful thinking.

The disk or nebula from which the Sun and planets formed has now vanished, reminding one of the Cheshire Cat in Alice in Wonderland that disappeared, leaving only its smile behind. Disks like our solar nebula begin when fragments break off one of the molecular clouds. What causes the clouds to break up? What determines the size, rotation and spin of the fragments? Why do some bits finish up as double stars? Shock waves coming from supernovae have been a popular way to break up dense gas clouds. However, the clouds perhaps simply collapse under their own gravity once they become cool enough.

One might also confidently have expected the primitive solar nebula, a fragment from a molecular cloud, to be nicely uniform in composition, and indeed this used to be the common perception. This topic has had an interesting history. It reflects a common approach to many scientific problems. What we cannot see or measure, we imagine as simple and homogeneous. This approach is related to the human tendency to underrate what we do not understand. This intellectual safety valve allows us to live with the unknown, and is no doubt responsible for the popularity of religions, astrology, mysticism and the other engaging fantasies that the human mind has dreamed up in the absence of facts. These provide nice stories, like flat Earths that are obvious to children. Reality is a bit more complex.

There used to be two competing models for the original size of our solar nebula. The large models for the primitive nebula contained about double the mass of the Sun. In these models, half the mass vanished into the Sun; the planets were made by breaking up the disk into a number of fragments which condensed into giant puffballs, called giant gaseous protoplanets. This model had the agreeable feature of forming the giant planets more or less instantly during the brief lifetime of the nebula. But the evidence for such large disks has vanished, for the disks observed, as in the Orion nebula, are mostly around 10% of the mass of the Sun.

The lower limit to the original size of our disk is obviously given by the present masses of the Sun and planets. One must add in some extra gas to compensate for that lost from the region of the inner rocky planets and the hole at the asteroid belt. In such models, the nebula behaves very differently from a massive one. It did not break up into planetary-size bits of its own accord. Instead, the dusty grains of rock and ice separated from the gas and settled towards the center plane of the disk. Boulder-size bodies grew by sticking the particles together, and larger chunks formed by collisions. Eventually, quite large bodies, the size of mountains, arose. These looked like our asteroids, and are called planetesimals. The reason for using this name is that the current model for building planets, at least in our solar system, is known as the planetesimal hypothesis. The term has a respectable history, dating back to the beginning of last century.

A critical question is, how long do these disks survive? The time from the initial separation of the disk of gas and dust from the molecular cloud out in the galaxy, to the point where the proto-star is large enough to ignite the nuclear furnace, is somewhere between a hundred thousand and one million years. Once the star begins to shine, the gas that did not fall into the star is soon driven out from the inner nebula, in times perhaps as brief as one million years. Disks of debris remain in the inner reaches of the disk and persist for a long time. Many old stars are surrounded by such disks that may represent analogs of our asteroid belt or Kuiper Belt.

Our inner planets are rocky. Gas is now present far away from the Sun at Jupiter. The satellites out there have plenty of water ice, except for the special case of Io which is cooked by being so close to Jupiter. Closer in, we have samples from the region between Jupiter and Mars. These are the meteorites that come from the asteroid belt. They date back to the beginning of the solar system and have a special significance since they tell us about temperatures at that remote epoch. Out at about 3 AU, it was just hot enough to melt ice. The meteorites and terrestrial planets that were a little closer to the Sun have lost not only water and other ices, but varying amounts of elements like lead, potassium and other easily vaporized elements.

Clearly our nebula was not an inert disk, but a dynamic and turbulent system with much energy to dissipate. Observations of disks around other stars reveal just how much complexity is out there. Few are symmetrical. Most naturally formed disks, like hurricanes or spiral nebulae, are not nicely symmetrical. A large spiral storm system on Earth, cyclone, hurricane or typhoon, according to taste and local folklore, is a good parallel. In these storm systems, there are a lot of thunderstorms and local turbulence within the overall large spiral structure. They are probably a reasonable model for conditions in the early disk that surrounded the Sun.

The lifetime of gas in the nebula is crucial for the formation of planets. If too much gas is left, planets will migrate quickly and fall into the star. To get circular and coplanar orbits requires just the right amount. It's a Goldilocks problem and the lifetime of the gas is a bit like the three bears' porridge [7].

A dilemma: the clearing of the inner solar nebula

> "It remains enigmatic how volatile depleted parent bodies were formed in the solar nebula" [8].

We fortunately know the composition of the dusty component of the original nebula from the composition of the CI meteorites (Figure 3). One might have thought that the rocky planets were made of this dust. However these bodies differ from the composition of the CI meteorites (and of the Sun). The inner nebula, out to about 3 AU, where Mercury, Venus, Earth, Mars and the nearer asteroids reside, has lost not only gases and ices, but also many of the volatile elements. Thus about 60–80% of the original complement of potassium is missing from the planets. We know this because uranium, thorium and a rare isotope of potassium emit gamma rays and so are easy to measure in samples or from spacecraft. Potassium is a volatile element but U and Th are both refractory. Thus the K/U or K/Th ratio gives us a measure of the loss of volatile K relative to refractory U or Th. The initial solar

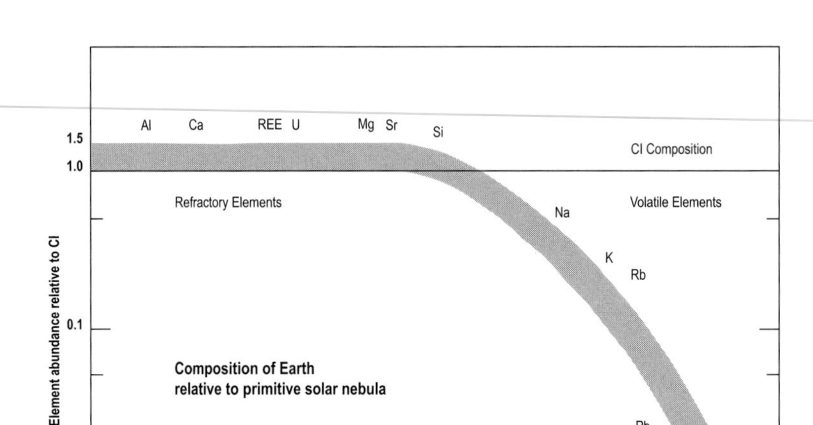

FIGURE 5 The composition of the silicate (rocky) mantle of the Earth relative to that of the primitive solar nebula. The figure shows the depletion of the "volatile" elements and some enrichment of "refractory" elements in the Earth. The "siderophile" elements that enter metal phases are not plotted, as many reside in the metallic core.

nebula value of the K/U ratio, as given by the CI meteorites, is near 60,000, but the ratio for the Earth is around 10,000, while that for Venus is similar within rather wide limits. The Martian K/U ratio is higher, near 20,000 and Mercury, significantly, has roughly similar values (Figure 4).

It has been well established that this depletion is a bulk planetary and not a surficial effect. Potassium, uranium and thorium, although distinct in chemistry, ionic radius and valency, are concentrated together in residual melts (and so into crusts). These diverse elements thus remain linked during planetary differentiation.

Many of the other volatile elements in the Periodic Table are also depleted, independently of the size of the planet. This depletion is well illustrated by the composition of the Earth, plotted relative to CI (Figure 5).

The striking feature of this plot, which resembles the abundances of the elements in the other terrestrial planets, as well in many classes of meteorites (except CI), is that the depletion of the elements correlates with volatility, not with any other chemical properties.

One alternative suggestion proposes that potassium is buried in planetary cores, but many other elements have to be buried as well and the depletion does not scale with planet size. Volatile elements are missing even in small stony meteorites. Another model, where potassium and the other volatile elements are evaporated during planetary accretion, is untenable as there is little evidence of the high temperatures needed, as discussed below.

We even have, courtesy of our remarkable meteorite samples, a date for this event. This is arguably their greatest contribution to our knowledge of the early solar system. The meteorites contain elements with radioactive isotopes that provide clocks that can be read. Thus some isotopes of lead, a "volatile" element to the planetary chemists, are produced from the radioactive decay of refractory uranium and thorium. Another example is the volatile element rubidium, a sister element that is inseparable from potassium, which has one isotope that undergoes radioactive decay to a refractory isotope of strontium.

Isotopic evidence from lunar samples and meteorites tells us that this volatile element loss occurred close to T_{zero} and was not due to later processes, such as evaporation. Reading these radioactive clocks tells us that there was a major separation of volatiles from refractory elements in the inner solar system right back at the beginning of the solar system, for which the agreed date is 4567 million years ago.

Now that we know when the loss of elements occurred, what caused it? Early notions were that the inner solar nebula was hot, a plausible notion as the Sun was close. In this model, the volatile elements were lost either because they did not condense as the disk cooled, or were evaporated in the hot nebula.

Some evidence from meteorites seemed to support this notion of a hot early nebula. Meteorites contain a few percent of grains composed of very refractory minerals. They are known as refractory inclusions or calcium-aluminum inclusions, or CAIs in common usage. They preserve a record of extreme temperatures and were formed when bits of dust were caught up in magnetic cycles near to the early Sun, being heated to temperatures around 2000 K. They suffered many cycles of evaporation and condensation before being flung far out into cool regions of the nebula, where they were incorporated into meteorites. Some grains even made it as far out as the icy home of the comets. The presence of these high temperature refractory minerals in meteorites immediately looked like evidence that the nebula had been hot and that these minerals had formed as it cooled.

But they only record special conditions very close to the Sun, not in the rest of the inner nebula. The concept of a cool nebula was reinforced by measurements of the temperatures in disks around other stars by astronomers. Their data show that the disks are cool, reaching a maximum of only a few hundred Kelvin at best, not the 2000 K needed.

The depletion of the volatile elements in the inner nebula clearly has something to do with activity from the young Sun. The truly decisive evidence came from workers in Chicago, who showed that there is no isotopic evidence for high temperatures in the inner nebula. They measured the isotopic ratios for potassium, a volatile element that is easily vaporized (Figure 6).

During heating experiments, the lighter isotope (mass 39) of potassium is lost relative to the heavier (mass 41) (it is the intermediate rare isotope with mass 40 that is radioactive). Although potassium is strongly depleted in the planets, the Moon and the asteroid Vesta, as well as in many meteorites, the ratio of the light and heavy isotopes, which is easily changed by temperature, remains constant.

So how does one remove volatile elements from a cool inner nebula? This is the dilemma. The answer lies not in the meteorites, but in space in the interstellar medium. This is the material that

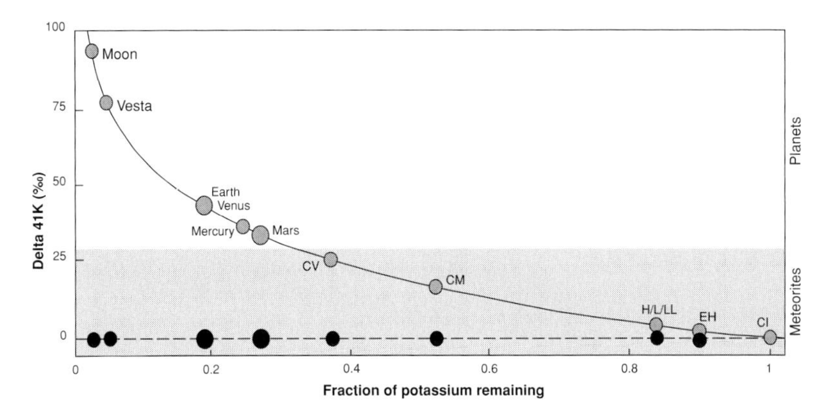

FIGURE 6 All the inner solar system bodies have lost potassium in varying amounts relative to its abundance in the primitive solar nebula (given by CI). The depletion of potassium in the various bodies is marked along the bottom (X) axis. Thus the Earth and Venus have lost about 80% of the original amount of potassium in the primitive solar nebula.

The curve with the planets and meteorites marked on it shows the enrichment that would be expected in the heavy potassium isotope (mass 41) relative to the lighter isotope (mass 39; data expressed in the standard format on the Y-axis as delta 41K in parts per thousand) if the loss of the element were caused by condensation from a hot nebula or by evaporation. The measured isotope ratios, dark circles on the lower line, show, however, that no change has occurred in any meteorite or planet. (EH, H, L, LL CM and CV are classes of stony meteorites). (Adapted from Humayun, M. and Clayton, R. *Geochim. Cosmochim. Acta* Vol. **59**, p. 2131, 1995).

flows in to form the Sun and we have some information on its composition. When the interstellar medium is observed, we can only see and measure the elements present in the gas (Figure 7).

The refractory elements are apparently missing and we see only the more volatile elements. Why is this so? The answer is simple enough. The refractory elements are present all right, but are locked up in condensed grains that we cannot see. We can observe with the current technology only the gas that contains the volatile elements. As the Sun gets large enough for nuclear burning to begin and it lights up, strong winds begin to sweep out the inner reaches of the nebula. The gas is mostly swept away by these strong out-flowing winds.

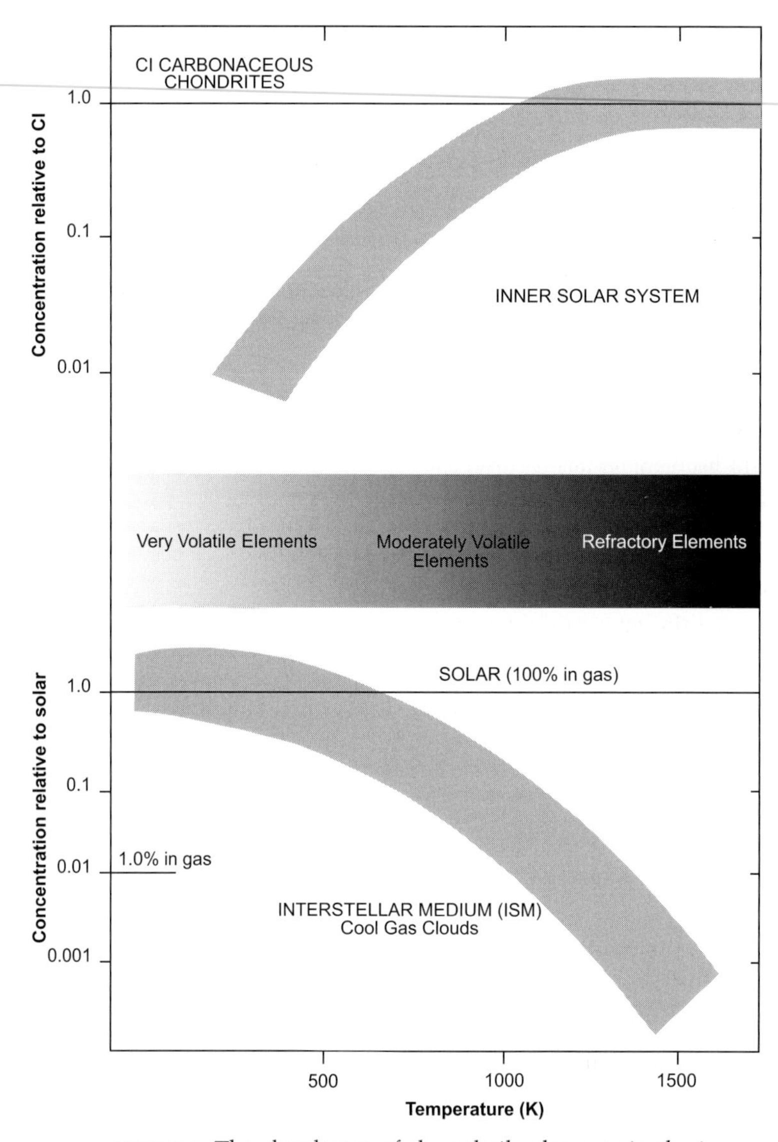

FIGURE 7 The abundances of the volatile elements in the inner solar nebula show an inverse relation to their concentrations in the interstellar medium. See discussion in text. (Adapted from Qing-zhu Yin in Chondrites and the Protoplanetary Disk. *Astron. Soc. Pacific Conf. Ser.* Vol. **341**, pp. 632–644, 2005).

They sweep away the potassium, lead and other elements that were in the gas, as well as the ices that include methane, ammonia and water. Thus the region where the terrestrial planets eventually form becomes deficient, not only in those elements, but also in water, carbon and nitrogen that were in the ices. Only the solid grains survive, so that the abundances of elements in the inner solar nebula look like the reverse pattern that we observe in the interstellar medium (Figure 7).

A formidable objection by geochemists to this model is that this process would lead to separation of radioactive element parents from daughters in the interstellar medium (ISM). Thus, refractory uranium would be contained in grains and volatile lead in the gas. The reverse would hold for volatile rubidium and its refractory daughter, strontium. This separation might produce exotic isotopic ratios and very old apparent ages that are not observed in our solar system samples.

But the interstellar medium is not a quiet place where such isotopic differences might be preserved. It is a violent and turbulent environment. The Giant Molecular Clouds have very short life spans, a few million years at best. Gas and grains are reformed on short timescales. It is far from any kind of equilibrium, beset by shock waves from supernovae, cosmic and X-radiation, stellar winds and magnetic fields. So the grains are subject to much sputtering and evaporation, and have short lives so that isotopic difference will be rehomogenized.

So in summary, the inner parts of the nebula were cleared of the gas, ices and depleted in volatile elements very early, probably within about 1 million years of the formation of the Sun. As the nuclear furnace ignited, strong winds blew outward from the Sun and swept away any remaining gas and ice in the inner disk. A few bodies, ranging in size from boulders to small mountains, survived the very strong stellar winds. We are standing on some of it. The finer material (dust, smoke, ice and gas) was swept away. Along with the gas went varying amounts of the volatile elements that had not condensed or found refuge in solid minerals.

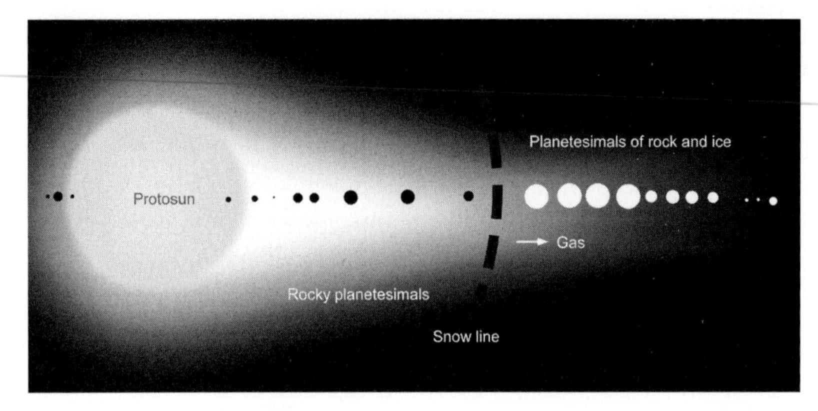

FIGURE 8 Snow line

As the early Sun turned on its nuclear furnace, strong stellar winds swept water and other volatile material out to between 3 and 5 AU, where the water condensed as ice and piled up in a snow line. This increased the density of the nebula at this location and so enabled icy cores about ten times the mass of the Earth to grow quickly. Two of these large cores then captured some of the gas that was also being driven away from the violent early Sun and became the gas giants, Jupiter and Saturn. The cores of Uranus and Neptune, farther out, only managed to catch a little gas. The terrestrial planets formed from the dry rock rubble left sunwards of the snow line. (Adapted from a slide courtesy of John Wood.) See also color plates section.

What happened to the ices and dust that was swept out of the inner nebula? The ices and some dust condense at the snow line (Figure 8) and were incorporated into the cores of the giant planets. Large amounts of ice have indeed been detected in the outer regions of a dusty disk around a young star. In this region the density of the nebula increases to a point that large cores of ten or more Earth masses form. If this process is fast enough, the cores catch the fleeing gas enabling gas giants like Jupiter to form.

But no distant reservoir of volatile elements exists. The amount of dust containing the volatile elements is trivial and is lost among the much more abundant ices, so there is no sink full of potassium or anything else out there [9].

REMOTE MATTER

Ben Jonson (1573–1637) in his play *The Alchemist* wrote with great insight that it was "absurd to think that Nature in the Earth bred gold, perfect i' the instant: something went before. There must be remote matter" [10]. We are now well informed about remote matter. One of the more impressive achievements of science in the twentieth century has been an explanation of the origin of the chemical elements. Hydrogen, helium, deuterium and a trace of lithium were formed in the Big Bang. The heavier elements, which the astronomers persist in lumping together as metals, were made in an immense number of stars over many billions of years and were scattered as condensed grains out into interstellar space. But these elements are distributed in a patchy fashion and some interstellar regions remain poor in them.

The two main sources are red giants, which shed a lot of mass out into space and the great stellar explosions of massive stars that we see as supernovae. During such stupendous events, mineral grains were formed in the shock fronts from the explosions radiated outwards.

What is remarkable is that some interstellar grains formed there are now found in meteorites. These grains have survived the rigors of interstellar travel from the supernova or red giant to the solar system, on a scale that would astound the crew of Star Trek. They remain to tell the tale of their distant origin, before the solar system began. Once here, they were caught up in the turbulent events in the early nebula. However, their presence tells us that the nebula was never heated enough to homogenize them.

They are identified by variations in the abundances of isotopes that are bizarre by the standards of the solar system. By this means, diamond and silicon carbide and other minerals that formed long before the solar system, have been identified in meteorites. The diamonds are tiny, typically containing only 25 atoms or so. As Ed Anders (b. 1928), who discovered them about 35 years ago while working at the University of Chicago, remarked, they are the right size to form the stones if bacteria wore engagement rings. These diamonds formed on the outskirts of supernova explosions, although the exact process remains unclear. The silicon carbide was made in red giant stars as they swelled up and, like a celestial stripper, shed their outer envelopes.

3 Forming planets

> "Making planets is an inherently messy business. A growing planetary system resembles an overly energetic infant learning to eat cereal with a spoon. Some is consumed, but much of it ends up on the floor, walls and ceiling" [1].

Planets were curiosities in most ancient models of the universe. But the discovery of Uranus and later Neptune led to speculation about their origin. Clearly they were distinct from stars and so might form in a different manner. But such notions came very late once it was established that they orbited the Sun and that the stars were remote.

The problem of building planets is fundamental to the entire question of the origin of planets. Historically, this latter question has frequently been considered to have been solved, but the wide variety of explanations and solutions that have been offered, from the creation myths of primitive societies to the more recent, but numerous scientific attempts, have generally collapsed when faced with new information. The latest problems stem from the bewildering variety of exoplanets. There are several difficulties.

The principal problem is that chaos and stochastic events dominate planetary formation, well displayed by the results from the Kepler and Corot missions. Stars, in contrast, are comparatively well behaved, their spheres of gas amenable to the laws of astrophysics. So theorists attempting to extract some general principles that would govern the formation and distribution of planets face difficulties that might have daunted Hercules.

Another dilemma is that the planetary scientist, like the historian, had until recently, only one example, our planetary system, to tell the tale of former events. One must of course be skeptical about relics. There is a long history of fraudulent relics such as Piltdown Man and human footprints superimposed to dinosaur tracks.

Statistical treatment has been successful in dealing with the formation of stars, of which there is a great abundance. We are beginning to get enough samples of exoplanets that we can move away from the statistics of a single system, dangerous because improbable events can always happen once. This is usually illustrated by the tale of the only elephant in the Leningrad (now St Petersburg) Zoo. During the siege of the city by the German Army in the Second World War, they deployed a giant gun, capable of firing over 20 kilometers into the center of the city. The first shell hit the zoo and killed the elephant. Other versions blame the first bomb to fall on the city [2]. But the complex systems that are now being discovered reveal equally extraordinary, unpredicted and improbable arrangements of planets.

From necessity, this section is heavily biased towards our own planetary system, the only one in which we can currently observe the chemistry and physics that accompanies the formation of planets. Theories, however, are based ultimately on observations (*pace* Plato). Few predictions have survived unscathed. The abundance and variety of exoplanets now enables some tests of the Standard Model for our system. They illustrate just how restrictive were our ideas about the formation and distribution of planets based on our solar system. The term "Standard Model" implies more certainty than was perhaps intended. Planets form by chaotic processes, difficult to constrain within the limits of mathematical models.

THE COLLAPSE OF CLOCKWORK SOLAR SYSTEMS

Attempts to find general models for forming planets have generally been misled by three properties of our solar system. Firstly, Bode's Rule led people to anticipate that there might be a similar spacing of planets around other stars. Next, exoplanets were expected to be in circular co-planar orbits about their parent stars, a legacy from

Laplace's concept of the nebula. Our whole system effectively lies in one plane, with most of the bodies orbiting the Sun and rotating in the same direction. This was what had impressed Laplace and it used to be thought typical of planetary systems.

Then planets might also have been expected at an early stage to have settled into stable orbits and not be subject to wandering about. While small rocky planets, such as Mercury, might be found surviving high temperatures close to their star, gas giants were expected to reside at suitably remote distances.

The success of Newtonian mechanics in predicting the presence of Neptune led to the concept of heavenly clockwork driving our system. The well-known image of a man peering past a veil of stars to discover a sort of grandfather clock mechanism beyond, typifies this viewpoint. The clockmaker presumably lurked behind the clockwork.

Even so distinguished and skeptical a scientist as Lord Kelvin (1824–1907) was led to make the following comment: "There may in reality be nothing more of mystery or of difficulty in the automatic progress of the solar system from cold matter diffused through space, to its present manifest order and beauty, lighted and warmed by its brilliant sun, than there is to the winding up of a clock, and letting it go till it stops. A watch spring is much farther beyond our understanding than is a gaseous nebula"[3].

Bode's Rule seemed to give regularity and order to our system, probably accounting for its popularity. The geometrical progression in the spacing of our planets looked like part of a grand design. But the famous "rule", on which so much intellectual energy was expended, turned out to be a minor result of tidal forces between planets. It is not a property of fundamental significance, nor does it form part of a grand blueprint for constructing planetary systems.

All these notions were based on examination of our own system, but its apparent orderly nature was misleading. What observers mostly failed to note was the extreme diversity of the objects making up our planetary system. A clear look at our planets and satellites showed that all were different and looked like the end result of some random collection. No two planets or satellites are alike. This might have warned us not to expect too many similar planets or systems elsewhere.

Views of a tidy and well-organized solar system have not survived. As in so many other fields, ranging from astronomy to genetics, we are reluctantly compelled to realize that we inhabit a universe in which chance events play a major role. Such ideas of the randomness of nature are unpopular. They run contrary to the egocentric philosophies in which *Homo sapiens* occupies a central role in the universe and in which all is apparently designed for his comfort and well-being.

The reality is different. The universe is just another physical system in which random events occur. This recognition of the importance of chance events has been one of the more profound changes in our perception of the world since the construction of a clockwork solar system could be ascribed to a divine watchmaker.

Clockwork models of the solar system have now become irrelevant if charming curiosities, like orrereys.

WHY DO PLANETS FORM AT ALL?

Stars form from cores collapsing under gravity from rotating clouds of interstellar gas that have separated from molecular clouds. These systems are spinning and have excess angular momentum, that product of mass and radius. Angular momentum cannot disappear; it is "conserved", in the jargon, so that the gas spins out into a rotating disk as mass accretes to the growing star. Material does not fall directly into the forming star, but comes in though spiral waves in the disk.

These rotating disks are the sites where planets may form, as happened in our system. Angular momentum is retained. As matter falls into the star, angular momentum is transferred outwards. The consequence, as has been long known, is that our planets possess 99% of the angular momentum of the solar system. In contrast, the Sun rotates very slowly, about once a month. This process of forming a star is not very efficient, leaving over some material in the disk from which planets may form.

Disks may fragment into two or more pieces that collapse into two, or more rarely, three stars. But even these individuals are likely to have excess angular momentum so that planets may form from the disks around the individual stars.

But nothing is pre-ordained. Exoplanets have been observed around stars that form one partner of a binary system. Some planets form around tiny brown dwarfs. Much seems to depend on the mass of the star and its complement of "metals". Clearly stars have to form before the formation of planets becomes possible; planets are late arrivals on the stage. But the formation of a disk does not guarantee the formation of planets.

Unlike stars, making planetary systems is chaotic and wasteful. In our system, two gas giants and two ice giants dominated the scene, leaving rocky rubble in the inner nebula. The icy bodies that escaped the sweep-up into the giants were dispersed into the outer reaches, where they reside in the Kuiper Belt or the Oort Cloud. How unique this arrangement is remains a current matter for enquiry.

COMPUTER MODELS

> "We can anticipate an "ultimate" planetary formation model... incorporating detailed physics as well as being able to reproduce... planet populations (mass, radius, and orbital characteristics, including period) [that] will enable a deeper understanding of planet formation and migration" [4].

It remains a conceit of modelers that, given a large enough computer and the appropriate data input, a complete model of the solar system would emerge, a notion dating back to the early days of computers. This is perhaps driven by their ability to model evolutionary paths for stars. But stars are very different beasts from planets. The processes surrounding the formation of planets are essentially chaotic, impossible to predict any more than who will win the lottery, or indeed which horse will win the Kentucky Derby. Indeed, more recent simulations suggest that "analogs to our solar system do not appear to be common" [5], something that is becoming apparent as more exoplanets are discovered in unpredicted places.

THE NATURE OF PLANETS

Stars, although they vary in mass, are mostly constituted of gas (96% for the most metal-rich) and so are amenable to mathematical and physical treatment, a feature that has led to the thriving field of astrophysics. But there is a fundamental difference between stars and planets. Stars form "top-down" by condensation, essentially of hydrogen and helium gas, from dense cores in molecular clouds. Their major differences in mass, luminosity and surface temperature are well displayed on the celebrated Hertzsprung–Russell diagram (Figure 1) which is nearly a century old.

Planets, in contrast with stars, were assembled randomly, "bottom-up" from leftover material in the nebular disk, at least in our solar system, but likely elsewhere. Our planets are all distinct, forming from a complex mixture that can be loosely labeled as gases, ices and rock. From our observations both of our own and exoplanets, such bodies may form from any combination of these three components (Figure 9). There is no equivalent of the Hertzsprung–Russell diagram for planets, or much sign of one appearing.

In our solar system, we have eight planets, all of them distinct from one another in mass, density, composition, obliquity and rotation rates. Their only common properties are near-circular orbits and low inclinations to the plane of the ecliptic (the Earth–Sun plane), characteristics that enabled Laplace to conclude that they had originated from a rotating disk of gas and dust, the solar nebula.

While we still have only one planetary system to examine closely, it includes over 160 satellites but of these, none resembles another, even among the "regular" satellites. Like the planets,

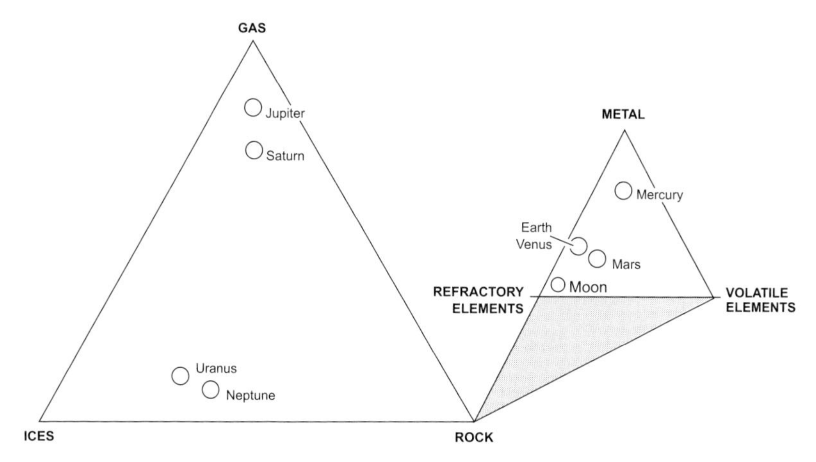

FIGURE 9 Planets may form from any combination of gases, ices and dust or "rock". The rocky planets, likewise, may show wide variations between metal content and those of the volatile and refractory elements. The approximate locations of our planets are shown. Eventually, exoplanets will fill all the space in these triangles. The planets also vary widely in mass, not displayed on this figure.

each satellite exhibits some peculiarities of composition or orbital behavior. This tells us that there is no uniformity in the processes of planetary or satellite formation from the gases, ices and rocky components of the primordial nebula. Clones of the planets of our solar system are consequently expected to be rare.

Our limited sampling of exoplanets displays an extraordinary diversity from our own system, in terms of variety, mass and spacing of planets while, to add additional complexity, many of these newly discovered planets are in highly elliptical orbits. It appears likely that we will find planets forming from Keplerian disks around young stars, which will occupy all possible niches available within the limits imposed by the cosmochemical abundances of the elements and the laws of physics and chemistry (Figure 9). Eventually, one expects the two triangles in Figure 9 to be filled.

There are only two basic ways to make things. The first is to start with something large and break it into smaller pieces. The second is to build up from smaller bits. An example of the first process is carving a statue out of a piece of marble; of the second, building a house from bricks. Such thoughts have resulted in two contrasting models for building planets. The first process breaks up the gaseous solar nebula, from which planets condense in the same way as stars. The second process, now generally referred to as the Standard Model for our planetary system, builds planets up out of smaller bits and pieces.

Planets may form from any combination of gases, ices and rock. These are the three components of disks surrounding young stars. While stars are basically gas with a dash (a few percent at most) of heavy elements, planets may be expected to vary widely in the relative proportions of gas, ices and rock. The relative amounts of the various ices may vary, while the rock fraction in turn may differ in the amounts of volatile and refractory elements as well as metal (mostly iron).

Examples of what has now become a truism abound in the results from the Kepler mission. Thus Kepler 10B, the "first rocky planet" discovered, is 4.6 times the mass of the Earth, with a density 1.6 times that of the Earth, or close to 9 gm/cm^3. Even allowing for self-compression in such a massive body, this could imply a large content of metal, in this case probably mostly iron. As the planet rotates in 20 hours around its parent star, the surface temperature is estimated to be 1800 °C, so perhaps the surface is molten.

Gas giants dominated by hydrogen and helium were expected to be common simply because gas dominates in disks. Rocky planets, although they may be abundant in number, form a minor part of planetary mass. In our system they account for about two Earth masses, while the gas and ice giants total over 445 Earth masses. But there are so many factors involved in forming planets that such simplistic calculations are probably over-ridden by other factors, such as the relative lives of the gas, ice and dust components in the circum-stellar disks. The difficulty in forming cores quickly enough to capture gas, or the rapid dispersion of the gas, may account for the many examples of Neptune-like exoplanets.

Planetary formation, at least in our system, is a very wasteful and inefficient process. Much of the nebular material either finished up in the Sun or was thrown away into outer space. What would one think of a factory that threw away most of the raw material? So little material finishes up in the rocky planets that an accountant might write off such an operation.

The old idea that all our planets formed more or less in place, as in a Laplacian or Neo-Laplacian model with planets condensing out of rings in the nebula, has given way to the notion of migration. Planets are mobile. This has been reinforced by the discovery of gas giants or Hot Jupiters, residing next to their star. No one has proposed a viable model for their formation so close to a star. But this early migration can only occur while gas is still around in the disk. In the absence of gas, later migration is driven by interactions with other planets.

In our system, Jupiter initially moved inwards, depleting the asteroid belt while the other giants moved outwards, pushing the leftover planetesimals out into the Kuiper Belt and the Oort Cloud.

However, the solar system analog does not fit the new planetary systems that we have found, for no two seem alike. They show many differences among themselves and none looks remotely like our system. This new information from these very limited examples raises the possibility of the uncommon nature of our own system.

There is no sign of design, intelligent or otherwise. Planetary-forming processes in our system seem to be essentially accidental and repetition of the particular sequence of events in another system seems as unlikely as multiple wins in a lottery.

If planets share any common factor, it is uniqueness.

The mass of the central star is crucial for the survival of a planetary system. High-mass stars have short lives, extending from 30 million years for stars of nine solar masses to only 3 million years for blue giants of 120 solar masses. The lifetime of the primitive nebula is another critical factor. Too short a life might not give time for solids to coagulate into planetesimals in the inner nebula. Too long a life might enable gas to collect around planetary cores and so a gas-rich planet might even form in the inner nebula. Infrared observations indicate that disks are dispersed on brief timescales with estimates ranging from 3 to 6 million years. By this time, the dust has coagulated into planetesimals. The gas had probably gone earlier. Gas loss, either by collapse on to Jupiter-like bodies or driven away by early stellar winds, seems to be decoupled from dust loss. In summary, the mass, angular momentum and lifetime of the disk of dust and gas are all random factors that together determine the final outcome.

Although it is frequently asserted that the discovery of Hot Jupiters or Hot Neptunes and the like require new models for making planets, such comments fail to understand that planetary formation is essentially stochastic. The processes of forming planets are chaotic, with all outcomes possible. Seeking one model is an illusion. This is in startling contrast to stars, amenable to the equations of astrophysics.

THE TIMING OF PLANETARY FORMATION

This subject suffers from falling between the studies of meteorites and astrophysical models of the nebula. Although gas made up over 98% of the original solar nebula, we are standing on a rocky planet that has only a trace left. When did this drastic decline happen? There is strong evidence that young stars lose their disks of gas within a few million years following the birth of the star. This presents us with an interesting problem. Our gas giants that are currently far from the Sun must have formed while the gas was still around. In great contrast, the Earth and the other inner planets were put together after the gas was gone.

So we begin to get some clues about when our rocky planets formed. The gas-rich planets had to form early within a few million years of the formation of the Sun. The inner rocky planets were assembled later from the leftover bits and pieces after the gas was gone.

The ultimate significance of the planetesimal hypothesis for our present discussion is that accretion from large precursor bodies is a random process. The formation of rocky planets is not an inevitable consequence of the accretion of planetesimals, but depends on the early history of the nebula. Before rocky planets can form, the gases and ices must be driven away. Although such events are presumably common around young stars, the presence of Hot Jupiters in close stellar orbits may indicate that timing is crucial and it seems unlikely that the sequence of events that occurred here will be repeated in detail elsewhere.

Finally, many of the volatile elements (notably carbon, but including nitrogen, phosphorus, sulfur, potassium, sodium and copper) that are essential to life as it occurs on the Earth, were mostly lost from the inner nebula. There is thus a certain irony in the fact that, at least in our solar system, these elements that are essential for life, are depleted in the habitable zone.

Snow lines

As the early Sun turned on its nuclear furnace, strong stellar winds drove water and other volatile material out to around 3 to 5 AU, where the water condensed as ice and piled up in a snow line (Figure 8). This increased the density of the nebula at this location and so enabled icy cores at least ten times the mass of the Earth to form quickly. These cores could then capture some of the gas that was also being driven away from the violent early Sun. The result in our system was the rapid growth of massive Jupiter, less massive Saturn with two cores, Uranus and Neptune, mostly missing out by their cores forming further away or perhaps a little later.

Giant planets

The most fundamental division in our planetary system is the difference between the giant planets and the small terrestrial planets. But even the giant planets differ significantly among themselves. Jupiter and Saturn, in addition to their massive gaseous envelopes, possess cores of rock and ice that are between 10 and 15 Earth-masses. In contrast, Uranus and Neptune, which are 14 and 17 Earth-masses respectively, contain only a few Earth-masses of gas and are mostly composed of ice and rock. These ice giants are thus analogs for the cores of Jupiter and Saturn. Jupiter and Saturn differ from the ice giants merely because they captured much larger amounts of gas. Significantly, Neptune analogs are very common among the exoplanets, so that nature seems to build cores readily. Gas giants are less common, so capturing gas seems a more chancy affair.

In addition to the distinction in composition between the giants and the terrestrial planets, there is also a major contrast in mass. Mercury, Venus, Earth, Moon and Mars contain only a trivial amount (two Earth-masses of rock) compared with the total of 445 Earth-masses of gas, ices and rock that reside in the giant planets.

It was only after the Sun turned on its nuclear furnace that strong solar winds developed, sweeping out the inner nebula, with the ices condensing at the snow line. The formation of any of our planets was thus a very late event in the history of the disk; beginning only after the Sun was essentially formed [5]. Gas giants are expected to be uncommon around low mass stars, where the disks are expected to be poorer in gas and perhaps to dissipate more rapidly. Very massive planets are more likely around massive stars, just as larger parents tend to produce larger children. But doubtless surprises await.

The lifetime of disks is only a few millions of years, between 3 and 6 million years, so that Jupiter and Saturn had to acquire their complement of gas within that period [6]. The early growth of these massive cores enabled them to begin capturing the gas (H and He) before the nebula was dispersed. Jupiter was able to accrete about 300 Earth-masses of gas. This is much less than that present in the original nebula, with the result that Jupiter does not have the composition of the Sun, but is enriched in the ice and rock component, or metals, by a factor between 3 and 13 [6]. Saturn, with a similar size core, managed to capture only about 80 Earth-masses of gas and is so more strongly "non-solar" in composition. Uranus and Neptune lost out almost completely and finished up with minor amounts of gas.

These non-solar compositions of the giant planets are key evidence for their "bottom-up" or core accretion models of formation from the solar nebula. The core accretion model indeed faces some problems of timing relative to the lifetimes of nebulae, although the times required to form the cores and collapse the gas on to them are not well constrained in the models and probably can be fitted into the few million years of disk lifetimes.

The alternative model for giant planet formation, by condensation directly from the gaseous nebula, is usually referred to as the disk instability model. Its main attraction is fast formation (giants form within a few thousand years), but it also faces theoretical difficulties. Although disks may break up, whether giant planets form from these clumps remains uncertain.

Apart from this, there are two serious objections. First, the giant planets are predicted to be of solar composition, but our giants are enriched by several times in ices and rock relative to the Sun.

Saturn, Uranus, Neptune and Jupiter all possess metal-rich cores. In the core accretion model, these form near the snow line and accrete their gas envelopes later, after they grow to about 10 Earth-masses. In the disk instability model the cores settle out after the giant has condensed. Such cores rain out due to differences in density once a planet has formed. For example, the metallic core of the Earth collapses due to the density difference between molten iron and silicate. But the temperatures and pressures within the early Earth are much lower compared with a giant like Jupiter. In the center of a giant, such density contrasts between different materials do not exist. The interior of a giant like Jupiter has pressures of 50–70 million bars, with temperatures up to 20,000 K. The material at these high temperatures and pressures is present as a plasma of protons and electrons, so-called "degenerate matter". A core cannot rain out in such a giant gaseous protoplanet in the manner that the iron core of the Earth forms. So it seems that the core has to be present to begin with, around which the gas can subsequently accrete.

The exoplanets provide additional evidence in support of the core-accretion model. Some are gas giants like Jupiter, but they form predominantly around stars with high metal contents. So nature apparently needs metals (i.e. ices and rock) to build gas giants elsewhere. This correlation between the existence of exoplanets and metal-rich stars is prima facie evidence that giant gas-rich planets form around ice and rock cores.

The most massive planetary systems are indeed found around stars that are both massive and metal-rich. But this merely reiterates one theme of this book. Any conceivable planet may form from gas, ices or rock, constrained only by the cosmochemical abundances of the elements and the laws of physics. Conversely, lower-mass systems found so far tend to be around stars with low mass and that are metal-poor. Although this is consistent with the core accretion model, one must beware of poor statistics and selection effects.

Finally, the existence of Uranus and Neptune,which are mostly ice and rock with about 1 or 2 Earth-masses of gas, shows that nature indeed managed to make two cores within our own system. Although the proponents of the disk instability model account for these ice giants by evaporating gas from larger bodies due to the intervention of another star, such ad hoc explanations do not explain how cores came to be there in the first place. Gas is difficult to lose even at high temperatures, as the presence of Hot Jupiters shows. Strong radiation or eroding winds from a nearby star would also produce chaos in the Kuiper Belt and the Oort Cloud, but no such effects are apparent.

Clearly the Hot Jupiters have not formed in place. No model allows for the amount of gas required to survive so close to a star. So migration definitely takes place. If too much gas is present, planets will migrate into their star. If there is too little, orbits may not be circular or indeed coplanar. Our solar system seems to have solved this particular Goldilocks problem. But models for the formation of gas giants beyond the snow line predict an inner nebula from which the gas was mostly gone. Clearly there are plenty of little understood problems about how planets migrate. Indeed it is surprising that planets exist at all, as current theories predict that migration rates are higher than is reasonable for planetary survival [7].

Perhaps the Hot Jupiters are driven in by gravitational interactions with giants further out. But it is interesting that those multiple systems that have a lot of smaller planets close in to their star do not seem to have giants nearby. This raises again the question of whether small inner planets would survive the passage of a Jupiter through their midst.

Some giants are found far away from their stars, again in positions where the amount of material in the disk was too meager for them to form either by core accretion or collapse. Presumably, like colonists, they have migrated to these distant locations by interplanet scattering, just as their cousins, the Hot Jupiters, migrated inwards [8]. Such improbable locations, coupled with the prevalence of eccentric orbits for planets, illustrate the violent dynamical environment in which planets form.

THE LIMITS TO GROWTH

Why did Jupiter stop growing? What were the limits to the growth of the giant planets? Why are they not larger? Why did not all the material in the nebula finish up in one giant planet, so that the system would resemble a double star? The answer is that Jupiter, having taken everything within reach, cleared a gap in the nebula and ran out of material within its grasp.

For the ice giants, Uranus and Neptune, the answer is also clear. They formed too late to catch much gas before it was all gone. Thus the formation of planets is self-limiting: they run out of material. If the nebula had been bigger, our space traveler would have seen yet another pair of double stars, instead of a single star surrounded by eight planets and some odds and ends. He would have noted its unusual character, for he would have seen other planetary systems that were wildly different from our own.

ROCKY PLANETS AND THE PLANETESIMAL HYPOTHESIS

> "In the context of planetary formation, impact is the most fundamental process" [9].

Following the formation of the gas and ice giants, all that was left in the inner nebula was dry rocky rubble, of which the inner belt asteroids are analogs. There is little evidence of the presence of either gases or ices during the accretion of the Earth. Rare gases are notably depleted in the Earth; their abundances are trivial compared to those in the nebula. Indeed, the relative depletion of the lighter ones (e.g. neon) suggests that they were derived later from carbonaceous chondrites. If the Earth had accreted in a gas-rich nebula, one would expect that ices would have been present and that the Earth would have accreted much more water ice, as well as methane and ammonia ices. In this case, the planetary budget of water, carbon and nitrogen would be orders of magnitude greater than observed, so we might have been a "water planet", with incalculable consequences.

Thus in contrast to the giant planets, the inner planets accumulated from the depleted debris after the gaseous and icy components of the nebula had left. It is worth noting that elements such as potassium and lead, which are much less volatile than water, are scarce in the Earth, while the primary minerals of most meteorites are anhydrous.

The material in the inner nebula, beginning with grains, grew into bodies that ranged in size from meter-sized lumps to Moon-sized bodies, before colliding to form the terrestrial planets. These building blocks are termed planetesimals, a term that originated in a somewhat different sense with T. C. Chamberlin (1843–1928) and F. R. Moulton (1872–1952). The best surviving analogs are the inner belt asteroids, along with Phobos and Deimos, the tiny moons of Mars.

The asteroids supply us with meteorites that enable us to decipher the early history of the solar system. We would be hard pressed indeed to sort out what happened when without them. The planetesimals were dry, volatile-depleted and had wide variations both in the abundance and oxidation state of iron. These fluctuations for such a common element serve as an example of the complexity among the planetesimals.

Some bodies were differentiated into metallic cores and silicate mantles that are common in the bodies in the asteroid belt. The basaltic meteorites (eucrites) that are derived from the asteroid Vesta, 450 km in diameter, are examples of such early processes. These provide evidence of the eruption of basalts on the surface of that asteroid at about 4560 Myr, a date that is within a few million years of T_{zero} [10].

The chief consequence for the formation of the terrestrial planets is that many of these planetesimals melted and differentiated within a few million years of the origin of the solar system. The evidence from meteorites and from the zoned arrangement in the asteroid belt is that most asteroids sunwards of about 3 AU were melted. Our diverse meteorites come from many different parent bodies. Most ordinary chondrites contain reduced metal, sulfide and silicate phases. In addition, they have also been depleted in volatile elements, displaying similar geochemical fractionations to those observed in the terrestrial planets. The heat source for melting these small bodies was probably short-lived radioactive isotopes of aluminum or iron. The result is that the Earth and the inner planets were assembled from objects that had previously melted and differentiated.

The larger terrestrial planets, Earth and Venus, took much longer to form than their smaller relatives, Mars and Mercury. They were formed very early in a violent environment. Two stages may be distinguished in the process of forming terrestrial planets. The first is a fairly rapid build-up or runaway accretion of hundreds of bodies that may approach the Moon, Mercury and Mars in size. This occurs rapidly on timescales of a million years or less. Mercury and Mars represent survivors of the final population that accreted to Venus and the Earth. As the planetary embryos became larger, gravitational effects became dominant and massive collisions between planetesimals became the norm. It took somewhere close to 100 million years for this multitude of bodies to be assembled into the Earth and Venus. The assembly of the terrestrial planets is hierarchical; many of the objects accreting to the Earth were of lunar-size.

Giant impacts were thus rather common and varied from head-on collisions of the sort that produced Mercury, to mergers of planetesimals, mass loss and break-up of bodies through to glancing collisions. Finally, one at least the size of Mars, named Theia, which might have survived as a planet in its own right if it had not collided with the Earth, formed the Moon as a result of the glancing collision with our planet. This Moon-forming event was among the last of the giant collisions.

Computer simulations for our inner planets indicate that, before the final sweep-up there were probably over 100 objects about the mass of the Moon (1/81 Earth mass), ten with masses around that of Mercury (1/20 Earth-mass) while a few exceeded the mass of Mars (1/11 Earth-mass), most of which accreted to Venus and the Earth. In addition there were a multitude of km-sized planetesimals.

During the later stages of planetary accretion in an essentially chaotic environment, the large planetesimals were widely scattered, so that during the final accumulation of the Earth and Venus, much mixing from near and far took place. In contrast to the accretion of the smaller planetesimals from restricted zones, the material now in the Earth and Venus probably came from the entire inner solar system. But whether these two planets or two or three other alternatives were the final outcome, was a matter of chance.

The planetesimal hypothesis predicts the occurrence of very large collisions in the final stages of accretion. These account for many of the features of the inner planets and greatly influence early crustal development and evolution. The impact of a Mars-sized object accounts for the origin, angular momentum and composition of the Moon. The high iron/silicate ratio in Mercury can be accounted for by removing much of its silicate mantle and can be explained by the collision of Proto-Mercury with an object of about 20% of its mass.

Late veneers of icy material drifting back from near Jupiter added water and traces of other exotic elements.

Some of these massive collisions have sufficient energy to melt planets, thus enabling metallic cores to separate from the rocky mantle. Can we date the separation of planetesimals into metallic cores and silicate mantles? One seductive method employs the metallic element, tungsten, which prefers to enter the iron cores. It has a radioactive isotope that produces an isotope of the element hafnium which prefers to stay with silicate minerals in rocky mantles. This provides an excellent way to date the formation of cores and mantles. But the caveat is that this process is going on in many different planetesimals that eventually finish up in the planets. Massive cores and mantles of planetesimals were added to the Earth randomly. All contribute some tungsten and hafnium so that now it is difficult to know which event is being dated. So the interpretation of the popular tungsten-hafnium isotopic system remains controversial [11].

Was the process of forming metal cores efficient, or were some metals such as nickel, cobalt and platinum, left in the mantle? The question remains open. To add to the complexity, traces of these elements were added to the mantle as a "late veneer" after our core formed. Assembling rocky planets is a chancy affair.

The formation of rocky planets through collisions of smaller bodies also accounts for the variations in composition of the terrestrial planets, as the planets accumulated from planetesimals that had already undergone many collisions. Thus some diversity of composition can be expected. Early planetary atmospheres may also be removed or added by cataclysmic collisions, accounting for the significant differences among the atmospheres of the inner planets. The collisions occurring during accretion are quintessential stochastic events. Of course, the probability of impacts of bodies of the right mass and at the appropriate angle and velocity to produce the Moon or remove the mantle of Mercury is low. However, other collisions during the hierarchical accretion of the terrestrial planets involving different parameters might produce equally anomalous effects such as a Moon for Venus, no Moon for the Earth, or different masses, tilts or rotation rates for the inner planets.

The variations in composition and later evolution of the terrestrial planets are thus readily attributable to the random accumulation of planetesimals with varying compositions. Computer simulations indeed have difficulty in reproducing the final stages of accretion of the inner planets, commonly producing fewer planets with large eccentricities and wider spacings, that emphasize the importance of stochastic processes in planetary formation.

It was Victor Goldschmidt (1888–1947) who pointed out that the metallic sulfide and silicate phases in meteorites were analogs [12] for the core and silicate mantle of the Earth. Although this generalization still holds, it has not proven possible to correlate specific classes of meteorites, either alone or in combination, with the composition of the bulk Earth. So although the Earth has a general chondritic composition, it cannot be linked either to a specific meteorite class or some mixture of the many groups.

The asteroid belt constitutes not much more than 5% of the mass of the Moon, and so it is a poor quarry from which to get material to build the planets. This depleted state of the asteroid belt itself is due to the early formation of Jupiter and dates from the earliest stages of the solar nebula. Thus it predates the accretion of the Earth and the inner planets. Oxygen isotopes vary among the different classes of meteorites and show that no class matches the terrestrial data, except for the enstatite chondrites. But there are other problems between these meteorites and the Earth; it is a coincidence that the Earth and the enstatite chondrites share the same oxygen isotopic composition. As is well known to philosophers, similarity does not imply identity.

The random nature of inner planet formation

"Chaos is a major factor in planetary growth" [13].

The inner planets completed their accretion in an essentially dry and gas-free environment, as shown, among many other observations, by the extreme depletion of the rare gases, such as neon and krypton, in the Earth. The fact that many planetesimals were differentiated before they were accreted into planets provides another set of variables in the early solar system. The evidence from the meteorites of early metal and silicate segregation is suggestive; one has to make the bricks before one can build the house. If these bodies, formed of iron cores and silicate mantles, were smashed up and re-accreted in differing proportions, then substantial changes in metal/silicate ratios might have occurred, as happened with Mercury.

In spite of the variations displayed among our four rocky planets, would such processes, if repeated often enough, produce clones of the Earth in other systems? The satellite systems of the giant planets provide some insight. Thus although we have only one set of planets, three of the giant planets, Jupiter, Saturn and Uranus, possess substantial regular satellite systems, while the capture of Triton was probably responsible for the destruction of any primordial satellite system around Neptune.

Although these miniature solar systems around Jupiter, Saturn and Uranus might have been expected to be similar, they are all quite distinct and it is often remarked that they might just as well belong to separate systems. Thus the processes that formed regular satellite systems from disks around the giant planets, although doubtless similar, did not result in a uniform product. That a large element of chance has accompanied the formation of the satellite systems reinforces the inherent difficulties in constructing general theories that can reproduce the details of planetary systems.

This is illustrated by the variations among the four rocky planets themselves that have resulted from the accretion of planetesimals in the inner nebula. Mercury, only 5% of the mass of the Earth, is perhaps only one-quarter of the mass of the body that collided with the Earth to form the Moon. It forms an example of a body accidentally left over that has survived by reaching a stable orbit.

Like Mercury, Mars is a survivor that might just as readily have been swept up into a larger planet. It owes its small size to the depletion of its neighborhood by the earlier growth of Jupiter. Mars is thus a good example of the differences that can result in planets assembled from a suite of rocky planetesimals. Nevertheless, although Mars is only about 11% of the mass of the Earth, it is of interest here as the other potentially inhabitable planet in our system. The crust seems to be made of basaltic lava, like the terrestrial ocean floors, and there does not seem to be anything on Mars that resembles continental granitic material on the Earth. So it is most unlikely that there are any lost continents, plate tectonics, subduction or Mt. St. Helens on Mars, reinforcing the uniqueness of terrestrial geological events.

In the inner solar system, only one other rocky planet managed to grow to around the size of the Earth. Thus, Venus is the most significant planet in our planetary system in terms of the present enquiry, as it shows what happens when Nature tried to duplicate our planet. Moreover, it provides a good illustration that while planets may be similar, they are not necessarily identical. It is not only sufficient to make a suitable planet; the subsequent geological history is also a crucial factor.

Large collisions in the final stages of accretion, including the Moon-forming event, are likely to have removed any primitive atmosphere on the Earth. The present atmosphere and hydrosphere of the Earth appear to be entirely secondary in origin. The terrestrial water budget, although uncertain, probably constitutes less than 500 ppm of the mass of the Earth. This is less than 1/1000 of the water budget in the primitive nebula, an amount so small that it could be ignored, except that we are here on account of it. The possibility that such stochastic processes have been responsible for the atmospheric evolution of the inner planets again emphasizes the chances involved in the formation of a habitable planet.

In summary, Nature made four rocky planets in our system, each one so distinctive that they could just as well reside in separate planetary systems. The bottom line is that all the inner rocky planets are different and owe their particular composition to the operation of random accumulation processes.

Sources of nitrogen and carbon

Both the Earth and Titan in our system have substantial amounts of nitrogen in their atmospheres. Nitrogen and carbon are essential for life on Earth and, presumably, for life on any other habitable planets. Both elements are present in trace amounts on our rocky planets compared to their great nebular abundance. In the original nebula they were present as ices such as methane or ammonia. These ices were driven away very early on by the active Sun, and the rocky planets were assembled from dry, gas-free planetesimals, as discussed above.

Curiously, the source of nitrogen on the Earth is almost never discussed in the scientific literature. The source must be ammonia ice, delivered like water, by the drift back of a few icy planetesimals from near Jupiter. Carbon was probably delivered in an equally haphazard form, perhaps as methane ice or perhaps in the form of polycyclic aromatic hydrocarbons (or PAHs), that are common in the interstellar medium and in CI meteorites. Given the random nature of the delivery process, planets might acquire much carbon or nitrogen, very little or none at all. So it is just not sufficient for a planet to reside in the habitable zone. The elements needed for life have to arrive in a manner best described as haphazard.

CRUSTS: ADDING ICING TO THE CAKE

Once again, this section is strongly biased towards our own planetary system, currently, the only one with accessible crusts. Once a rocky terrestrial planet has been assembled, what happens next? The release of gravitational energy during the process of assembly from the infall of large planetesimals results in hot, mostly molten planets. Radioactivity provides an ongoing heat source after they have solidified. The energy derived from the residual heat from the collisions and that supplied by the radioactive elements drives the tectonic processes that result in the variety of planetary surface features that we observe.

It seems inevitable that rocky planets, like bakers, cannot resist making crusts, heat being the prime cause in both cases. Although trivial in volume relative to their parent planets, crusts often contain a major fraction of the planetary budget of elements, such as the heat-producing elements K, U and Th, as well as many other rare elements. Meanwhile, the familiar continental crust of the Earth on which most of us live is of unique importance to *Homo sapiens*. It was on this platform that the later stages of evolution occurred and thus enabled this enquiry to proceed [14].

Crusts have undeniable advantages for scientists: they are accessible. Unlike the other regions of planets that we wish to study, such as cores and mantles, you can walk on crusts, land spacecraft on them, collect samples from them, measure their surface compositions remotely, study photographs, or use radar to penetrate obscuring atmospheres. In spite of this accessibility, the problems both of sampling or observing crusts are non-trivial; most of our confusion in deciphering the history of crusts ultimately turns on our ability to sample them in an adequate fashion.

This advantage of relatively easy access to crusts is also offset by their distressing tendency to be complex, so that one may easily become lost in the detail, failing to see the forest for the trees. This is particularly true of the continental crust of the Earth which is sometimes heterogeneous on a scale of meters. One consequence of this myopia is that one sometimes encounters claims that extrapolate from a small region to produce a world-embracing model. The furor over whether or not there was an early granitic continental crust is a familiar example of the perils of extrapolation from a handful of zircon grains preserved in younger sedimentary rocks. As Charles Gillispie has remarked "the inherent difficulties of the science [geology], Lyell thought, had rendered it peculiarly susceptible to the interpretations of ancient miraclemongers and their modern successors" [15].

However, just as the planets themselves are not identical, not all crusts are equal. Three types of crusts may be distinguished on rocky planets, conveniently divided into Primary, Secondary and Tertiary.

Primary crusts are formed as a consequence of initial planetary differentiation, caused, for example, by melting during accretion.

Primary crusts form on short timescales, within less than 1 million years following planetary formation. They may contain high concentrations of incompatible elements, because primary crusts are derived from mostly molten planets and so are able to collect trace elements from large volumes of planetary mantles.

The highland crust of the Moon, which now constitutes 8 or 9% of the Moon, was produced directly following the formation of the Moon and forms the type example. The ancient heavily cratered crust of Mars in the southern highlands is another possible example.

Secondary crusts form on much longer timescales. These result from melting of the lower temperature fractions of silicate mantles of our rocky planets. This process produces various species of basalts. Typical examples include the oceanic crust of the Earth, the present surface of Venus, the volcanic outpourings that gave rise to the northern plains on Mars and the Tharsis plateau, as well as the dark lunar maria.

Likewise the oceanic crust of the Earth constitutes only 0.1% of the mass of the planet, but as it is continuously formed from partial melting in the mantle, the total volume extruded over geological time has perhaps amounted to 10% of that of the Earth. Because magmas that form secondary crusts are derived from partial melting of limited volumes of planetary mantles, secondary crusts are enriched in incompatible elements to a much lesser degree than primary crusts.

Tertiary crusts are formed by dehydration or melting of secondary crusts. The continental crust of the Earth remains the only current example in our solar system. Because such crusts build up over long periods of time, they can contain great enrichments of incompatible elements, due to the continuous recycling of the secondary crust by plate tectonics. It has taken the Earth over 4000 million years to form the continental crust. It seems to be difficult to produce such tertiary crusts, as no evidence has appeared elsewhere in the solar system of the existence of such silica-rich crusts. But our useful platform constitutes only a trivial fraction (0.4%) of the mass of the Earth, while the granitic upper crust produced by melting of the lower crust amounts to perhaps 0.1%.

In the absence of the recycling of the oceanic crust that occurs on the Earth, crustal growth on the other rocky planets in our solar system is essentially irreversible, a process that results mostly in surfaces covered with various species of basalt, the so-called stagnant-lid regime. On the Earth, the continental crust remains buoyant, destroyed only by erosion, unlike the oceanic crust that is rapidly recycled back into the mantle by plate tectonics.

The collisions of asteroids and comets with planets

> "Giant impacts appear to be a likely and normal consequence of planetary accretion" [16].

Once more, this discussion focuses on our system for obvious reasons. Our planets all circle the Sun in the same sense, a consequence of the original rotation of the disk of gas and dust from which the system formed. One might expect that if our planets formed from such a rotating disk, they might all be upright with zero obliquity. They might either spin at the same speed, or in some tidy mathematical sequence, just like the regular spacing of Bode's Rule. Although our planets mostly spin on their axes in the same sense, anticlockwise when viewed from above the North Pole, there are exceptions to this regular arrangement. Venus rotates slowly backwards, while Uranus lies on its side. Moreover, the planets are all tilted, and rotate at different speeds. All bear some signature of having experienced unique events. Why is this so?

The different tilts and spins of the planets constitute the best evidence for massive impacts in our early planetary system. No model involving condensation from a disk in an orderly manner can account for the rather untidy situation in which the planets now find themselves. It would be extraordinary if all planets were tilted to the same degree, just as much as if all planets were identical. If the planets showed zero tilts or some obvious regularity in their spins, it would be possible to entertain an orderly origin for them.

The 23.5° tilt of the Earth provides us with the seasons because of the variable amount of sunlight received over different areas. At present, Mars has a similar tilt to that of the Earth, but over time the red planet wobbles through 60°. Venus, in great contrast, has only a very small tilt. Opinions differ about the cause. Was Venus hit head-on and stopped in its tracks? Perhaps it never suffered a giant impact. The slow rotation of Venus may indeed be the primitive state for most planets and their varying spins, like their tilts, are due to massive collisions late during their formation.

The tilts of Jupiter and Saturn may be due to a combination of collisions and perhaps warps in the gaseous nebula. Jupiter has only a slight tilt, but Saturn is inclined at nearly 30° to the common plane of the solar system, more so even than the Earth. Uranus and Neptune have significant tilts. The most extreme case is Uranus. This planet is 14 times more massive than the Earth. It takes the impact of something the size of the Earth to knock a planet of this size over. Uranus has a collection of nine rings and 15 satellites that all rotate around its equator. These must all have formed after the planet rolled over. Thus our planets bear, not only on their battered faces, but also in their varying tilts and spins, silent witness to the trauma surrounding their birth.

Curiously enough, the plane in which our planets lie is tilted at 7° to the equator of the Sun. This is rarely discussed. Perhaps some late torque twisted the gaseous nebula away from the plane of the Sun's equator and is responsible for part of the tilt of the giant planets.

Both Mercury and the Moon represent special cases in the inner solar system. Both owe their unique character to the effect of large collisions. Like Mercury, but for a different reason, the Moon is an anomaly in the solar system. It has a very low density, in contrast to the high density of Mercury. One body has too little iron and the other has too much. Thus collisions can produce strange bedfellows.

Pluto and its very large satellite Charon, are not only in an eccentric and highly inclined orbit, but rotate around each other, at right angles to the rest of the system. Although such a curious situation is difficult to produce in an orderly system, it is the likely result of a large collision.

If large collisions are a characteristic feature of the final stages of the accumulation of the planets, then it is not possible to predict the details of the events. Such collisions occurred at all times and stages in the history of our planetary system, and perhaps elsewhere. This process began with the sticking together of grains in the primitive disk of dust and gas. It continued with the growth of bodies which eventually reached the size of small planets. Innumerable impacts occurred during the sweep-up of these smaller bodies into the planets. These culminated in the massive final collisions that tilted the planets and started them spinning at different rates. Because the biggest bodies grow last, the largest impacts occur towards the end of the process. Some collisions spun out disks from the giant planets from which satellites formed.

How do we know about the former existence of large bodies? These have now vanished. Because we live on a planet on which erosion removes craters rather quickly, the significance of impacts in solar system history has only slowly been appreciated. The lunar surface, visible through the smallest telescope or binoculars, is the classic example. Spacecraft photographs show that from Mercury out to the satellites of Uranus, a massive bombardment struck planets and satellites. Craters of all sizes are present. They range from micron-sized pits due to impact of tiny grains on lunar samples, up to giant ringed basins the size of France (Figure 10).

The extent of this early bombardment on the Moon is revealed by the presence of at least 80 basins with diameters greater than 300 kilometers. Another 10,000 craters are in the size range 30–300 kilometers. These formed before the Late Heavy Bombardment ceased, about 3850 million years ago, but the flux continues, albeit slowly.

The Late Heavy Bombardment, so well displayed on the Moon, where it is termed the Lunar Cataclysm, seems to be well established as having happened within a short time around 4000 My ago. It was

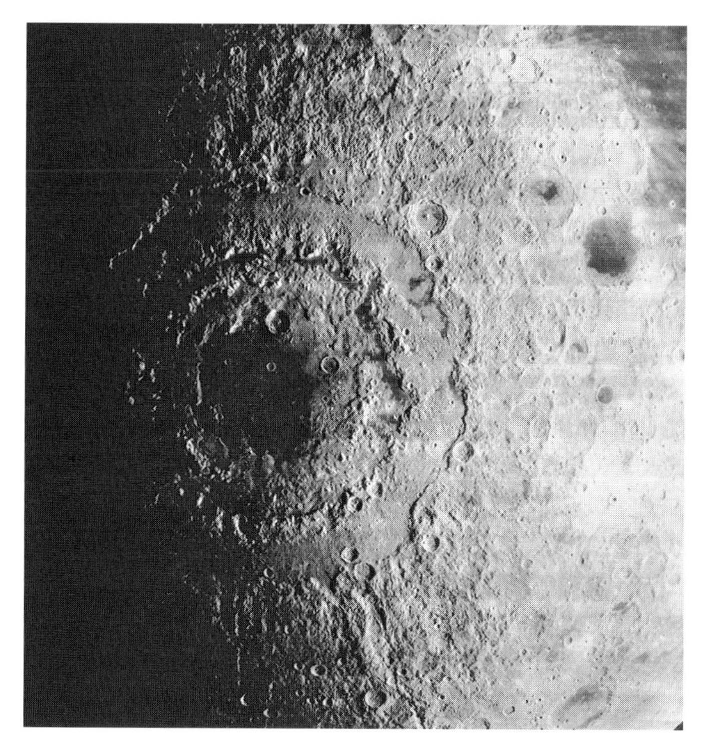

FIGURE 10 The site of a major collision on the Moon. Mare Orientale, the classical example of a multi-ring basin, about the size of France, was the site of a massive impact on the Moon 3850 million years ago, when a planetesimal about 50 kilometers in diameter slammed into the Moon and, in a few minutes, produced these concentric rings of mountains, several kilometers high. The central region has a thin veneer of basalt, erupted much later from deep in the mantle. The diameter of the outer mountain ring, Montes Cordillera, whose eastern edge is sometimes visible from the Earth, is 900 kilometers. The small dark area of mare basalt to the northeast of Orientale is Grimaldi, visible from the Earth through binoculars. The smooth dark area on the northeastern horizon is the western edge of the basalt plains of Oceanus Procellarum. (NASA Orbiter IV 187 M.)

caused by movement among the giant planets which sent a shower of asteroids and comets into the inner solar system [17]. This event needs to be clearly distinguished from the much earlier collisions that occurred during the formation of the planets themselves. The most dramatic of these that affected us was the collision of a Mars-sized body with the Earth that resulted in the formation of the Moon. But all of these impacts happened within 100 million years of the formation of the system. The Moon-forming impact was one of the last great ones, although smaller collisions have continued, trivial on the scale being discussed here.

In earlier times, it was thought that the craters on the Moon were mostly due to volcanoes. Some workers right up to the manned landings of the Apollo spacecraft held this view. The history of the debate over whether volcanoes or meteorites caused the craters on the Moon is another fascinating chapter in the history of science. It is full of misconceptions, misidentifications and faulty conclusions. Unfortunately I must leave most of this fascinating topic to the historians of science.

Among the many puzzles that the craters presented to students in previous times, the fact that the craters on the Moon were mostly circular, was one of the most difficult to explain. The reasoning ran that meteorites would hit the Moon at all angles, and so should produce mostly oval craters. It was not until this century, with its unfortunately close acquaintance with high explosives, that the cause of the circular nature of craters was understood. An instructive example of the formation and nature of craters was formed when, on July 1, 1916, a mine was exploded at La Boiselle, in a futile attempt to breach the German defenses on the Somme. The crater was 85 meters wide and 25 meters deep, complete with a five-meter-high rim and an ejecta blanket of chalk. It closely resembled one due to meteorite impact. There are many more recent examples.

Cosmic impacts are more destructive than our puny efforts. Thus, a rocky fragment 250 meters in diameter, around the size of a football stadium, hitting the Earth at perhaps 20 kilometers a second, has the explosive energy of a thousand megatons of TNT. No matter at what angle it hits the Earth, it will bury itself and explode like a bomb. The resulting explosion digs a circular crater 5 kilometers in diameter and about 1 kilometer deep. Within a few minutes, the surrounding countryside for several kilometers from the crater will be covered with a deep blanket of rubble and broken-up rock thrown out of the crater.

About every 20 million years, an asteroid in an Earth-crossing orbit hits the Earth and forms a crater 20 kilometers across. The impact of a larger Apollo or Aten asteroid colliding with the Earth would produce a major catastrophe, perhaps wiping out life on Earth. If life survived, evolution might go off in some other direction. Bizarre forms might appear again as evolution starts tinkering in a new direction. The extinction of the dinosaurs, and much else 65 million years ago at the end of the Cretaceous Period, was due to the impact of an asteroid about 10 kilometers in diameter. That devastating collision produced a crater in Mexico over 200 kilometers in diameter. The rise of mammals to fill the vacant ecological niches was a consequence.

The most spectacular landforms discovered by spacecraft are probably the giant impact basins such as Orientale on the Moon (Figure 10). However, like the immense Martian landscapes of Valles Marineris and Olympus Mons, that are so impressive when photographed from orbit, they would be less striking on the ground. Early workers, notably G. K. Gilbert (1843–1918), Ralph Baldwin (b. 1912) and Harold Urey, all drew attention to the circular nature of the Imbrium basin on the Moon. It's the size of Texas. They considered that it formed when a large asteroid hit the Moon. The later discovery of the Orientale basin provided a nearly perfect example of a ringed impact basin (Figure 10). With five concentric rings of mountains, it is like a great bulls-eye, 900 kilometers across. Photos of Mars and Mercury show similar immense basins surrounded by rings of mountains. Radar mapping of Venus showed that even that thick cloud-shrouded atmosphere provides no protection if the impacting bodies are large enough.

The objects that now hit the Earth are either comets coming from the outer reaches of the solar system, or asteroids and meteorites. These are perturbed from their orbits in the asteroid belt by collisions or the gravitational effects of Jupiter. The Earth also sweeps up occasional meteorites, knocked off from Mars or the Moon by large impacts. Bodies less than about 10 meters in diameter are burnt up in the atmosphere. Because of their high speed, they have energies equivalent to a few megatons of high explosive, such as TNT.

The cause of the event over Tunguska in Siberia on June 30, 1908 has been the subject of vast speculation. An explosion occurred at 7.30 am local time, at an altitude of 5 kilometers, and released energy equivalent to about 20 megatons of TNT. The resulting shock wave blew down the Siberian forest for over 1000 square kilometers. Although many bizarre explanations, including the arrival of some anti-matter, have been proposed to account for this sudden explosion, the reality is more prosaic. The cause is now known to be a stony meteorite colliding with the Earth. The meteorite was only the size of a small building, about 60 meters in diameter, but it was travelling at perhaps 20 kilometers a second. If the meteorite had exploded over New York, it would have completely demolished the city. Such an event happens somewhere on the Earth perhaps every 1000 years. Fortunately, the statistical chances of such a large meteorite hitting a great city occur only once in a million years.

Are such bombardments universal?

Probably. It all depends on chance. It has been argued that moons are common around Earth-mass planets, where they would be useful in stabilizing tilts. Large late collisions are probably frequent, although Venus in our system apparently escaped. But the parameters of our Moon-forming collision seem special enough that the event is judged a rarity.

Although it is argued here that our individual planets are the results of a sequence of chance events, the architecture of the solar system itself now appears to be due to a similar sequence of chance occurrences [17].

4 The exoplanets

"It must be admitted that Nature is more creative than the human imagination" [1].

Trying to write a current account about exoplanets reminds one of the definition of Post-Modernism: "it's so new that it's out of date already". Looking for exoplanets around stars used to be an unpopular topic, shunned by professional astronomers; but times change. One is reminded of the comment by Talleyrand that "treason is a matter of dates". Now a barely concealed excitement suffuses much of the serious scientific literature. Among many quotations the following give a flavor: "the search for extra-terrestrial planets – rocky worlds in orbit around stars other than the Sun – is one of humanity's most exciting science goals" [2], while the real goal seems to be "the detection and characterization of habitable worlds" [3].

The previous section showed what we have attempted to learn about the formation of planets, based strongly on observations of our own system. The discovery of the exoplanets has revealed many surprises and demonstrates once again the triumph of observations over theory. Although "one of the most surprising aspects of the hundreds of known exoplanets is their broad diversity" [2], this was less surprising to students of our own solar system, in which all the planets and satellites display an astonishing range of sizes and compositions. This results in the notorious difficulties of trying to classify or even define planets, exacerbated now that we have such a variety of exoplanets. The long established upper limit for planets of 12 or 13 Jupiter-masses, that marks the lower limit of deuterium burning, has now been extended to at least 25 Jupiter-masses into the brown dwarf desert, and will need to go to higher masses. Nevertheless and consistent with its name, the desert remains thinly populated. Below the 12 Jupiter-mass limit, problems also occur because of the discovery of free-floating objects down to at least 3 Jupiter-masses. Are these bodies that are not linked to any star, "free-floating planets" ejected from the nest, or "sub-brown dwarfs"? We seem to be approaching a classification for planets, brown dwarfs and the like based on where they formed. Truly a nightmare for classifiers.

THE LONG-AWAITED DISCOVERY

The search took over a century; claims going back to 1855 with many false leads mostly based on astrometry, which is the precise measurement of the positions of stars. The principles for discovering planets by their gravitational influence on the parent star were eventually understood, but all depended on reaching an exquisite level of precision. But even when this was accomplished, the search remained difficult. Thus in August 1995, the results of a 12-year survey found nothing, though it was using the highly precise techniques of the radial velocity method to search around 21 nearby main-sequence Sun-like stars.

More relevant was the report in 1992 of the first true exoplanets, three Earth-sized bodies, but interest soon faded as it was revealed that they were "dead worlds" in orbit around a pulsar and so not the long-sought El Dorado. Although the scientists involved might have felt that they had solved a problem that had been around for two millennia, their discovery was soon overwhelmed by a report from Geneva on October 6, 1995, of a Jupiter-mass body in an astonishing 4.2-day orbit around a Sun-like star, 51Pegasi, 50 light years distant. The finding was confirmed only 6 days later from California, although it took a little longer to dispel the doubters, who attributed the finding either to an intrinsic wobble of the star or to a binary star system, rather than to a planet in orbit.

Totally unpredicted by any theory of planetary formation, the finding of a gas giant within about 8 million km of its parent star where the temperature is 1200 K, was astounding. Our sun-baked Mercury, in contrast, is about seven times further away at about 58 million km (0.39 AU). A deluge of exoplanets, now numbering in the thousands, soon followed the discovery.

But the eclipse of the discovery of the pulsar planets reveals an uncomfortable truth. Few were interested in the inert Earth-mass planets that had been found in orbit around pulsars. These are rotating neutron stars. Any planets unfortunate enough to form from the debris disk following the supernova, of which the pulsar is the residue, are bathed in copious radiation from the pulsar. In addition, the pulsar provides little visible light, making these bodies an unlikely place for life either to arise or to flourish. Unfortunately for the discoverers, habitable planets were the real goal. The controversial problem that had been around for 2500 years remained with us. So interest in these pulsar planets waned and was displaced by the multitude of bodies that were discovered orbiting regular stars.

Why was this so? This question comes back to the real driver for supporting such expensive investigations. Funding them, like science in general, depends ultimately on popular support, more likely to occur in open secular societies. The Apollo Missions to the Moon accomplished a great deal of science, but were sent to our satellite for political, not scientific reasons.

The search now is not for extrasolar planets as such, but is revealed in the name of one of the principal promoters of the search, the SETI Institute, which is the acronym for "Search for Extra-Terrestrial Intelligence". It is the Little Green Men whom we are after, fuelled by the speculations of science fiction. The search may even be pursued on home computers that currently employ over 3 million participants. They are seeking radio signals from distant planets that one trusts will be intelligible if or when they are received.

So one must feel some sympathy for the discoverers of the pulsar planets who felt that they might have answered the ancient question. The real search is for life, assumed covertly or overtly to be intelligent. It is clearly enough stated as "A Search for Habitable Planets" which is the subtitle for the NASA Kepler mission. Like railway stations, planets, while interesting in themselves, merely provide a suitable platform for a larger purpose.

SEARCH TECHNIQUES [4]

The oldest was the simplest. If one can accurately determine the position in the sky of a star, a planet in orbit around it would cause a measurable shift. This method, known as astrometry, is the most obvious and is of respectable antiquity. It has the advantage that it can directly measure the orbits and the mass of the orbiting planet. But it has still to achieve enough precision to work from the Earth and no confirmed discoveries have yet been reported. Nevertheless, it is possible that a space mission could reach the required precision of measuring microarcseconds and so provide a catalog of Earth-mass planets near the Sun.

The most successful technique by far in looking for planets is the so-called Radial Velocity method. This is based on the Doppler principle of measuring the shift in the spectra as the star moves towards or away from the Earth under the gravitational influence of a planet. The stars and planets all rotate around their common center of mass, the so-called barycenter. Thus the presence of Jupiter in orbit causes the Sun to move its position at a rate of 13 meters per second (or nearly 30 mph). As the star moves towards the Earth, the spectral lines move minutely to the blue or shorter wavelengths. As the star recedes, the spectral lines move towards the red or longer wavelength of the spectrum. This is a miniscule example of the famous redshift that tells us that the galaxies are receding from us and that the universe is expanding.

The success of the method depends on exquisite calibration and temperature control to enable the tiny shifts in the spectral lines to be measured. This requires very high-resolution spectrometers, the best of which can detect movements of less than 1 meter per second. The key was to superimpose the light from the star on a terrestrial multi-line spectrum. Once these formidable technical problems were solved, hundreds of planets have since been discovered. The technique measures the orbital periods and the minimum mass of the planet, but the inclination of the orbit remains unknown. So the planetary lists discovered by this method are full of masses of the planets multiplied by sine i values (M sin i), commonly expressed relative to the mass of Jupiter. The inclination (i) of the planetary orbit is 0° if face-on and 90° if edge-on.

This intrinsic wobble of the star, termed "jitter", limits the radial velocity method, so that quiet stars tend to be selected. Older cool stars in the F, G, K and M classes (Figure 1) are among those with the least jitter and so such older slowly rotating stars are popular candidates. The K dwarfs are among the best candidates with the least jitter. The method is most useful when the star is quiet, allowing measurements of the effect of the planet on the movement of the star to be pursued below 1 meter per second. This level of precision allows the detection of planets down to a few Earth-masses. However, many stars are noisy at this level, a problem that is currently restricting the ability to detect Earth-mass planets. These cause a deflection of the star of around 10 cm per second. Doubtless, a technical fix will occur to improve the precision of the Doppler method by the necessary order of magnitude.

These biases need to be remembered in looking for correlations between or among planets and stars. Caution is however required, as the history of science is littered with false correlations. Among other successes of the radial velocity technique has been the discovery of many Neptune-mass planets or "super Earths" as well as many multiple systems. In these, sophisticated techniques are needed to sort out the contributions from several planets to the movement of the central star.

When the planet crosses in front of its parent star, so that we see the system edge-on with zero inclination, its true mass can be found. This forms the basis of the very useful Transit method which detects planets passing in front of their parent star. Such transits are an infrequent event, dependent on the observer just happening to be in the line of sight. However the method is precise enough that the dimming of the starlight (around 0.01 to 1%) can be measured, allowing the radius of the planet to be established. Its mass may then be determined by the radial velocity method. The transit method also provides useful data on planetary atmospheres as well as the size of the parent star. The technique is used in the successful Corot and Kepler Missions that operate in space.

The procedure is not without its share of problems, such as jitter and false detections from binary stars. More serious and unexpected are variations in luminosity of younger stars that require longer observing times. The Kepler Mission is currently surveying 156,000 stars. It has already detected over 2000 possible planets from near-Earth-size to larger than Jupiter; among its goals is to establish the abundance of rocky planets within the habitable zone. Over 50 of these, all larger than the Earth, have already been discovered as well as many multi-planet systems.

Another method, known as Gravitational Microlensing, uses the gravitational field of a star-planet system to act as a lens to magnify the light from a more distant star that happens to lie in our line of sight. Bending of light by gravity, as predicted by Einstein's Theory of General Relativity, was famously demonstrated in 1919. When a foreground star passes in front of a distant star, it acts as a lens, magnifying the light curve of the distant star. Any planet orbiting the foreground star shows up as a separate peak on the light curve.

There are some basic problems with the microlensing technique. Planetary detections are thousands of light years distant. The chances of finding a transiting planet are rare, less than one in a million at a particular moment, so that huge numbers of stars need to be surveyed. Finally the detections are a one-shot affair. So it is most useful for finding out how many planets are out there and hence gives us the best statistical sampling of planets. Although only a few detections were reported initially, the technique has achieved instant fame by discovering a multitude of free-floating objects that are either "planets" or "sub-brown dwarfs", according to taste.

Many other methods have been suggested or are being developed. One such is direct imaging. This requires blocking out both the light from the star and its scattered light in the detector, either with coronographs or interferometers. Already many of the formidable technical difficulties have been overcome and several large planets and brown dwarfs have been observed. The ultimate goal is to obtain spectra from Earth-mass planets that might indicate the presence of life, but daunting technological challenges remain, particularly in the development of the instruments that need to operate in space. The next required step seems to be orbiting instruments such as the proposed NASA Terrestrial Planet Finder or the ESA Darwin Mission that could directly inspect the planet. In this context, the proposed replacement for Hubble, the James Webb Telescope, may be less suitable.

The radial velocity and transit techniques currently remain the most successful, but selection effects limit the discoveries. Thus the radial velocity method is much better at finding high mass planets in close orbits, rather than low mass planets in wide orbits. Transit methods are obviously limited to seeing planets in edge-on orbits and, once again, large planets in close orbits are easier to find. Microlensing, in contrast, can find low mass planets in wide orbits. Although the method is sensitive enough to find large moons it suffers because the observation can rarely be repeated; it's a singular event. Direct imaging is currently limited to large bright planets far from their stars. However, technology is moving at an extraordinary pace and many of these limitations are being overcome.

WHAT HAVE WE FOUND?

> "In the hunt for planets around stars other than the Sun, astronomers' primary objective is to find a world teeming with life" [5].

A very large number of exoplanets, currently now numbering thousands and including many multiple systems, have been discovered. Free-floating objects, that may be planets or sub-brown dwarfs in the cosmos, may even outnumber the stars themselves, as the Kepler

Mission has found. Any attempt to list them is out of date before it appears in print. Perhaps it should be mentioned here that only direct imaging reveals the planets themselves. The other methods reveal the gravitational effects of the invisible bodies on their parent stars.

Although there are now enough planets to enable some statistical studies to begin, such attempts are limited by the selection effects inherent in all of the discovery methods. Nevertheless an extraordinary variety have appeared. Planetary masses range from near Earth-mass to giants exceeding 12 or 13 Jupiter-masses, to the discomfort of classifiers. Orbital periods range from less than a day to over 14 years. The Hot Jupiters that are in close orbit around their star receive more than 10,000 times more stellar flux than Jupiter. The Corot mission, designed to look for transiting planets, has perhaps found more than was bargained for: some bloated, dense or extremely dense Hot Jupiters and some super Earths were among the first discoveries. But more bizarre was the discovery of a few massive bodies inhabiting the brown dwarf desert.

The number of planets discovered in orbits far from their star will increase with time because it takes extended periods of observations to find them. Some planets are hundreds of AU distant; others as close as 2 or 3 million km to their star. Although most planets are in prograde orbits, about 25% of Hot Jupiters are in retrograde orbits, that is, counter to the spin direction of their star, raising considerable problems for students of planetary dynamics. Eccentricities range from circular orbits ($e =$ zero), mostly for planets near their star, to those with eccentricities greater than 0.9, close to the value for a hyperbola at 1.0.

The Kepler mission, employing the transit method, has found over 200 multiple systems, many more than were expected. Most contain two planets, but systems with three, four and rarely five and six have been found, although the numbers drop off dramatically above three.

One unexpected feature is the number of coplanar orbits, although their eccentricity is mostly unknown. More surprising is that most of these planets in these multiple systems are small, less than the mass of Neptune. Transiting giant planets, that might have disrupted such systems, seem rare.

None of this was expected and demonstrates once again the power of observation over theory. Most expected something looking like our solar system; Hot Jupiters were the first surprise and theorists are struggling to explain the unpredicted findings. One is reminded of the remark that "it is difficult to make predictions, especially about the future", a comment from folklore that has been ascribed to people as distinct as Yogi Berra, Albert Einstein, Mark Twain and Winston Churchill.

A BEWILDERING VARIETY

"Strange New Worlds" [6].

Hot Jupiters were the first ones discovered (pace the pulsar planets), being the easiest to find employing the radial velocity method. But complexities soon arose. Although transit methods revealed that they were indeed gas giants, no one could explain how a gas giant could form in such a hot environment. A consensus soon arose that they must have formed out at several AU at a snow line and migrated inwards, due perhaps to interaction with the disk or other giants. Perhaps they are the survivors that stopped in time.

Many explanations are current and these giants have provided a rich feast for theorists. Among other problems, some were inflated and others very dense. The denser ones presumably have large metal-rich cores while some of the obese (but not all) have been inflated by their closeness to the star. To add complexity, some are in retrograde orbits. Other gas giants reside at respectable distances, as do our Jupiter and Saturn, with some further out at over 100 AU, discovered mainly by microlensing. They must have been scattered to such distances.

No theory of the formation of planets has enough mass in a disk to form them in place, although formation by disk instability has been claimed.

Many Neptune-like planets have also been discovered, some again very close to their star. Other colder Neptunes, orbiting their stars at a few AU, are very common. They seem to be true Neptune analogs [8]. Many are intermediate in size between the Earth and Neptune, filling a gap that exists in our solar system.

Terrestrial or rocky planets are common, both as revealed by the Kepler mission, and expected from the strong statistical trend towards lower mass planets. Many come with labels such as "carbide" planets, "water worlds" or "ocean" planets, that exist at present only in the imagination of the observers. Perhaps it is revealing that the title of a recent NASA symposium was "Exploring Strange New Worlds".

How many stars have planets around them?

> "The planets detected so far represent only the tip of the iceberg (9%) of all the existing planets" [7].

This sounds like a simple question but belongs in the category of "how long is a piece of string"? Is it important or is it just another interesting statistic? It also depends on both the type of star and the planet, so that it is perhaps a rhetorical question. The formation of disks is a natural consequence of the formation of stars. So, potentially any star could spawn planets leading to estimates between 1% and 100%. But the formation of both stars and planets depends on many chance events. First is the initial angular momentum of the collapsing fragment of a molecular cloud. This controls whether a single or double star forms. Other parameters include the mass of the disk and how long it lasts before being dissipated. Additional factors involve the type of star. Planets are more common around metal-rich stars, so they are naturally chosen as targets. A quarter of stars that are much more metal-rich than the Sun appear to possess planets.

Fewer than 5% of those with a lower metal content than the Sun appear to have planets around them. But selection effects, depending on whether one is dealing with gas giants or gas-poor Neptunes, may bias this estimate.

The real number of interest is how many Earth-mass or hopefully Earth-like planets are in habitable zones around Sun-like stars. So those stars that are not too different from the Sun, the F, G and K classes (Figure 1), have mostly been surveyed. If the search is really for Earth-mass planets, many other factors enter. High mass stars (O, B and A classes) are less popular for searches, as they "age too quickly to support the development of complex Earth-type life" on planets, to quote one website. Selection effects currently play a role. Then it is easier to find gas giants in close orbits. Low-mass stars, although probably very common, are more difficult to spot; giants far from their star are equally difficult to find.

The initial data from the Kepler mission revealed that the number of stars with transiting planets is 0.8%, but microlensing techniques indicate that most stars might have planetary systems. Another estimate suggested that 12% of FGK stars (those most suitable to harbor habitable planets) have gas giants within 20 AU [8]. When the transit method was employed to look for Hot Jupiters in the globular cluster 47 Tucenae, none were found, as noted above. The stars in this globular cluster are very old and poor in metals. As planets apparently require metals to form, this negative result is not very surprising. A similar result was found for another 12 billion-year-old globular cluster, Omega Centauri.

A much more sophisticated survey, based on the many stars surveyed by the Kepler mission, and which accounts for the sizes of the planets as well, shows that under 1% of solar-type stars have Jupiter-sized or larger planets in orbit. Smaller planets are much more common, with Earth-sized planets expected around 30% of solar-type stars.

Hot Neptunes and Neptune-mass planets are very common. Recent surveys show that most Sun-like stars have at least one planet in orbit and it is possible that planets are ubiquitous around stars, as the gravitational microlensing method suggests.

The discussion here refers to planets bound in orbits, but microlensing surveys have revealed that unbound planets may be as common as stars [9].

So the question is complex and many answers are possible depending on how the question is phrased. The bottom line is that probably many stars have planetary systems. Earth-mass planets, perhaps in close orbits, are likely to be common. How many of these are in habitable zones and truly resemble the Earth as suitable locations for life is another question, which is the real motivation for this search and for this account.

Masses and orbits

More massive planets tend to occur around massive stars, although again selection effects may be involved. Planetary masses above eight times Jupiter-mass are rare but the numbers increase rapidly with lower masses, mimicking the Initial Mass Function of stars. Eventually a current observational limit around a few Earth-masses is reached (Figure 11). Probably the trend continues with numbers of planets increasing as their mass diminishes down to an Earth-mass, or perhaps to objects like Mercury or even Vesta.

There is a pile-up of planets with orbits around 3 days due to the Hot Jupiters. This is caused by the migration of giant planets and their ease of detection. Another peak occurs for orbits around 200 days, but the valley between may be due to observational biases.

The dilemma of eccentricity

> "The eccentric orbits of the known extrasolar giant planets provide evidence that most planet-forming environments undergo violent dynamical instabilities" [10].

The fact that most of the exoplanets are in eccentric orbits (Figure 12a,b) was one of many unexpected results from the early discoveries. Eccentricities often reach extreme values; a few reach values

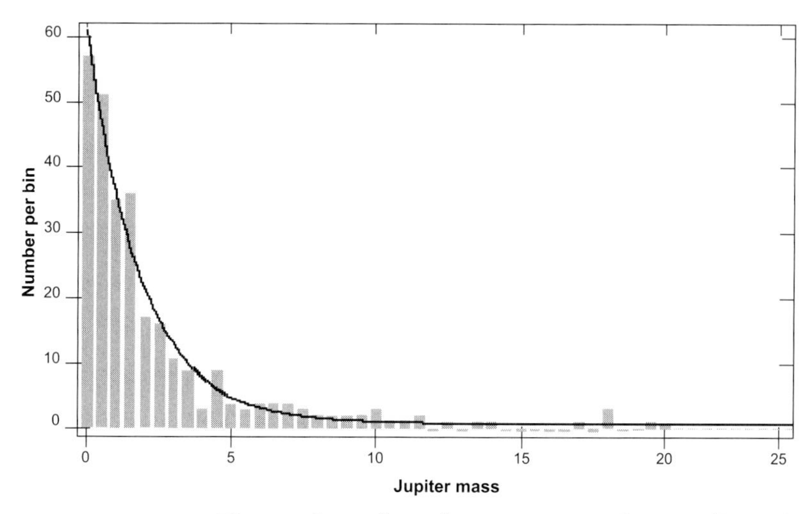

FIGURE 11 The numbers of exoplanets increases dramatically with smaller masses. Few are found with more than about 8 Jupiter-masses. A few stragglers extend into the "Brown Dwarf Desert" beyond 12 or 13 Jupiter-masses.

greater than 0.9. Beyond a value of 1.0 (a hyperbola), a planet would no longer be in orbit. One might expect that exoplanets that now reside close to their stars would have nearly circular orbits, due to the strong tidal influence of the massive star. Many do, but curiously not all of the orbits have been circularized although the planets are so close to the star that they must have been locked into synchronous rotation, always presenting the same face to the star. Some Hot Jupiters have eccentricities as high as 0.4. Perhaps this anomaly is due to the gravitational effects of other giant planets in eccentric orbits.

The eccentric orbits of the exoplanets resemble those of binary star systems. This caused some early confusion about whether we were indeed observing planets or binary star systems, but it is probably more of a coincidence than of any fundamental significance. What it is telling us is that circular orbits are difficult to achieve in nature. More massive planets seem to have more eccentric orbits than those of smaller mass, but the correlation is once again more suggestive than convincing.

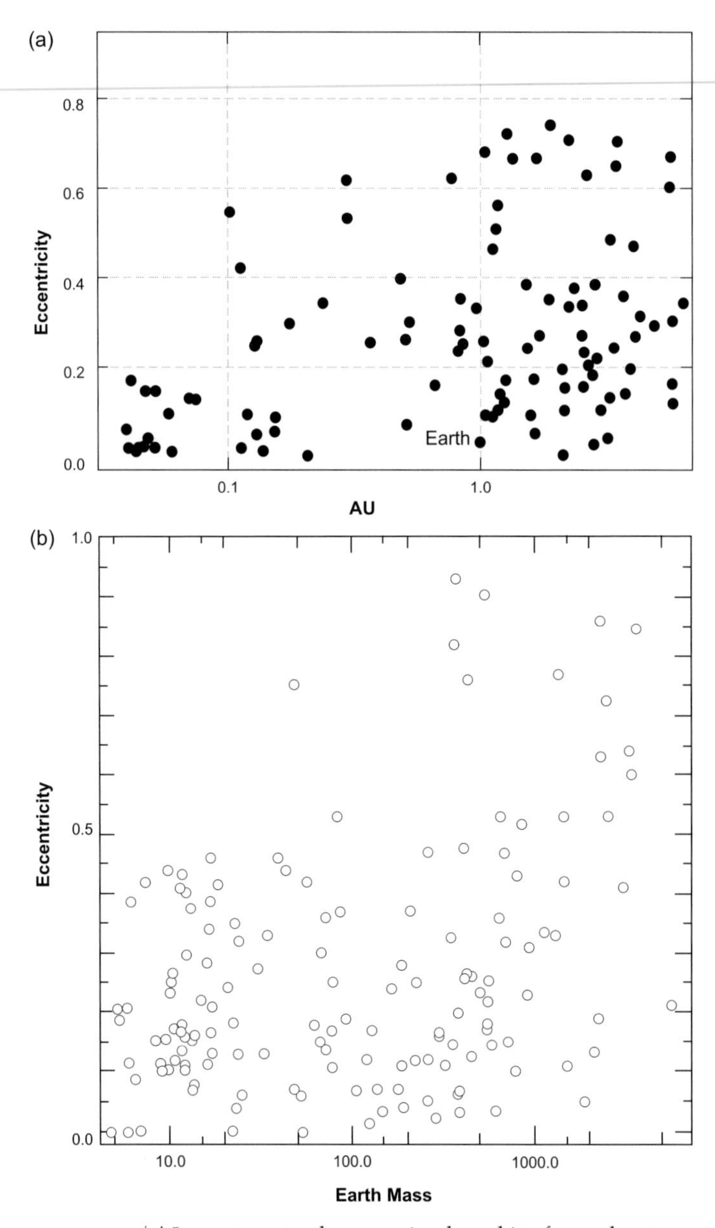

FIGURE 12 (a) In contrast to the near-circular orbit of our planets, exoplanets display a wide range of eccentric orbits, here plotted against distance from the star. The orbits of binary stars show a similar distribution of eccentricities. (b) A recent survey of lower mass planets, here plotted relative to Earth mass, shows a similar wide range of eccentric orbits (adapted from Mayor, M. *et al.*, *AA*, in press 2012).

Rocky inner planets seem unlikely to form in systems that have giants in highly eccentric orbits. This notion has been reinforced by the discovery by the Kepler mission that gas giants are rare in multiple systems in which inner planets are in coplanar orbits.

Most theories of planet formation begin with circular disks around stars from which planets form, much as Laplace had proposed. Eccentric orbits are thought to arise later from interactions with other planets, with the disk or even nearby stars. On reflection, this is perhaps the normal outcome from a system in which collisions are common. Some impacts might even result in the ejection of planets from the disk. If the free-floating objects turn out to have been formed in disks, then they may have been kicked out of the nest as a consequence.

Inclinations are mostly unknown for exoplanets, except for those planets observed to transit their star. Our planets, in contrast, seem well-behaved moving in their placid, nearly circular orbits that lie nearly in a plane of low or zero inclination (except for tiny Mercury with $i = 7°$).

The migration of planets [11]

Hot Jupiters were the first great puzzle to bewilder the planetary seekers. How could a Jupiter-mass body reside so close to a hot star. Planets in such close orbits are likely to be in synchronous orbits, with one hot face (>2000 K) gazing permanently at its star while the other is frigid (≈ 50 K). All current models tell us that gas giants form far from stars and no one has proposed a viable alternative. Giants form at snow lines, where dusty ice cores can form and collect gas to grow quickly before the gas is swept away. It used to be thought that our giants had not moved much. But appearances are deceptive and led to models that tried to form them in place. But it later became clear that even our giants had formed initially near the snow line, but had subsequently migrated to their present positions.

It also became clear that the Hot Jupiters had migrated inwards from cooler regions. Once started, migration inwards might seem difficult to stop, so all of the planets might have finished up in their star.

Clearly this has not happened. Current models suggest that planets migrating inwards are stranded when the disk runs out of gas, just as whales are stranded on beaches when the tide goes out. So the Hot Jupiters don't appear to be survivors, the "last of the Mohicans" [12]. But what might have happened to any small rocky planets in a habitable zone as the giants bulldozed their way through? Perhaps rocky planets might form much later?

Although these questions have formed a fertile field for theorists to explore, one must feel some sympathy for them in dealing with such unexpected and complex phenomena. Simple models of gas disks seem inadequate. Indeed, it seems that we are lucky to see any planets at all as "a major problem is that the effects of migration are typically predicted to be too strong to have allowed the formation of gas giant planets to complete" [11]. Many other factors such as turbulence, temperature and even magnetic fields are variable and difficult to account for. The very existence of planets and the final architecture of planetary systems depend on the unpredictable properties of the initial disk.

Significant migration also occurred in our system long after the gas-rich solar nebula had dissipated. The point that deserves emphasis here is that the present architecture of our solar system is the result of a sequence of random encounters. The system looks tidy and well organized into inner rocky planets, a belt of asteroids, four giants, a Kuiper Belt of icy planetesimals and an Oort Cloud. But this is the consequence of migration amongst the giants hundreds of millions of years after they had formed. Thus it should not be a cause for wonder that exoplanetary systems fail to match ours. The chaotic evolution of a planetary system is responsible for its final architecture.

Several different types of migration have been erected, apparently in an attempt to introduce some order into a process that appears driven by random factors. Type I migration occurs due to interaction with the gas in the disk. Such migration has to occur within the lifetime of the gas in the disk. This lasts only a few million years. Type II migration occurs a little after the giant planet has cleared a gap in the disk, but when some gas is still present. More likely processes are gravitational interactions among planets once they are formed. These may cause much scattering of planetesimals and later movement of giant planets, as seemed to have happened in our system.

But general explanations for migration are difficult to find, as the properties of the disks are so variable. Turbulence is important, but unpredictable. It is the familiar story of attempting to deal with stochastic processes in planetary formation. As a careful study of our system has revealed, random processes predominate.

The role of the star

Planets are known to form around all types of stars, including binary systems and brown dwarfs. This indeed might have been expected as the process of star forming itself forms a disk from which planets may form. Does the composition of the star influence the formation of planets? The short answer is, yes. Planets tend to form around metal-rich stars like the Sun. Stars with a lower metal content than the Sun form fewer planets; those richer than the Sun form more. This does indeed seem to be a genuine correlation, although there is an understandable tendency for planet hunters to target metal-rich stars.

The success of this relation between the composition of the star and the presence of planets has led to other attempts to find connections. Several other correlations have been proposed, but most are more suggestive than valid statistically. All suffer from incomplete sampling or selection effects. Massive planetary systems seem to occur only around stars that are both more metal-rich and more massive than the Sun. In contrast, low-mass systems are found around low-mass stars with low metal contents, but exceptions are known to these generalizations. But as though to confound modelers, a planet in a 4-day orbit, Corot 3b, with a mass 22 times that of Jupiter, has formed around an F-type star. This 2 billion-year-old star is similar to the Sun except that it is metal-poor.

But very massive stars have too short a life to be of much interest to seekers after habitable planets. It is usually suggested that there would be insufficient time for life to evolve very far on any planets that might form or where life might take root. Attention has thus focused on Sun-like or smaller stars (F, G and K classes, Figure 1) that live to a respectable age, giving time for life, not only to arise and evolve, but hopefully to develop both intelligence and radio transmitters.

It has also been suggested that stars that form planets have low abundances of lithium, but this has been shown to be yet another false lead. The element, lithium, is easily consumed during nuclear reactions and so its abundance naturally decreases with the age of the star. No statistical difference has been observed for lithium content between stars with or without planets [13].

Another idea suggests that stars with high carbon contents might form rocky planets made of carbides, rather than silicates. Perhaps this happens somewhere, but this notion ignores the fact that no matter what the carbon content of the star, during planet formation carbon is present in the disks mostly in the form of ices (e.g. methane). These ices are swept out of the inner parts of the disk, that is, the zone in which rocky or terrestrial planets form. So rocky planets in which carbides replace silicates seem likely rarities.

THE METALLICITY CORRELATION

Although stars are dominantly composed of hydrogen and helium, they may contain up to 4% of metals. The Sun has 1.4%. The ratio Fe/H is commonly used as a measure of "metallicity". This is due to the ease of determining the abundant element, iron, from the analysis of the spectra of stars in which iron lines are abundant. The use of Fe/H serves well enough as a measure of metallicity, but other "metals" such as carbon, oxygen, nitrogen, neon, silicon and calcium are formed separately from the iron group by the alpha process of element formation in stars. Thus their abundance is not closely tied to that of iron. However, this has only a minor effect on the correlation except for more primitive stars.

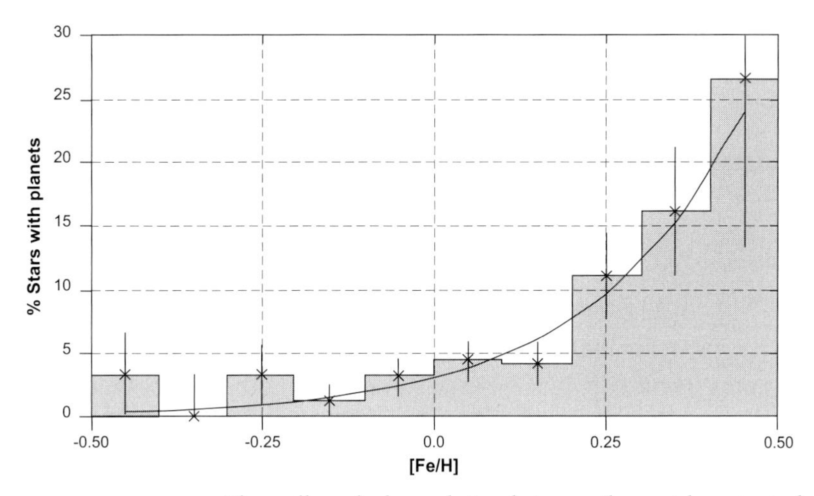

FIGURE 13 The well-marked correlation between the metal content of the parent star and the number of exoplanets. Fe/H is a measure of the metallicity of the star, as explained in the text, here plotted on a logarithmic scale relative to our Sun = zero (0.50 is three times the metal content of the Sun and −0.50 is one third).

There is a clear and striking preference of exoplanets to form around metal-rich stars (Figure 13). This was apparent from the earliest days of exoplanet discovery. This correlation of gas giants with metal-rich stars has proven robust and is not a selection effect.

Is the measured metal content an inherent property of the star? It was suggested at first that the metal-rich nature of stars harboring planets was caused by the infall of planets into the star, coating the atmosphere with metals. Apart from its logical inconsistency, this model was quickly shown to be inadequate and there is now a consensus that the high metal content is a fundamental property of the star.

Exoplanets that resemble Jupiter are common enough. Why should such gas giants need metals? If they form by collapse from a disk, the composition of the star does not matter. But if gas giants need a core of metals around which they collect the gas, their abundance around metal-rich stars becomes understandable. Indeed, the core accretion model for forming planets like Jupiter begins by forming a 10 Earth-mass core of ices and rock. The clear implication for models is that gas giants need metals to form.

Another random factor that enters is the amount of dust in the original disk. It has been suggested that if the gas/dust ratio is high, gas giants may form preferentially. In contrast, a gas-poor disk may form Neptune-type planets, which are mostly cores without much gas. But such effects may also be due to the rate at which the gas is driven away, and on the location of the core relative to the snow line. It seems likely that this correlation may be another selection effect. It should be recalled that both gas and ice giants exist in our own system, where the gas/dust ratio was the same.

In contrast to this well-defined relation between the metal content of stars and the presence of planets, many of the other claimed relationships among exoplanets seem much less convincing. Few of them would satisfy a statistician. It seems more likely that there is a continuum of masses, sizes, eccentricities, orbits, orbital distances and composition. So, little regularity exists among the planets, in great contrast to stars.

Multiple planetary systems

"Planetary systems configured like our own are rare indeed" [14].

The discovery of multiple planet systems, likely to resemble the solar system, was long expected. What was found was nonetheless just as surprising as the discovery of the Hot Jupiters. The multiple systems might have been expected to lie in a common plane, like our planets. But many systems do not, perhaps due to later collisions and scattering. Indeed, gravitational interactions of planets among themselves seem to be common. Several have finished up in so-called resonant orbits that are related by simple whole numbers (e.g. 1/2). This effect, although unexpected, is likely due to tidal interactions and migration once the planets have formed. Many of the systems are full or packed so that there are no stable vacant orbits between the planets.

Currently about 20% of planetary systems are multiple; this percentage is bound to increase. The Kepler mission has located many transiting multiple systems. One solar-type star (HD 10180) has five Neptune-mass planets (ranging from 12 to 25 Earth-masses) packed closely around it [15], with another two possible companions. This system is interesting as the star is older than 4 billion years, similar to our Sun. But how do several Neptune-mass planets arise and migrate inside the orbit of our Mercury? Neptune-mass planets are expected to arise in regions of enhanced density near the snow line far from the inner nebula. To add more to the puzzle, no gas giant that might have perturbed the system and driven the Neptune-like bodies inwards seems to be present.

Another distinctive system lies around Kepler 18, a Sun-like star. Three planets are in close orbits (3, 7 and 14 days). The outer two are Neptune analogs and the inner is a super Earth, with 6.9 Earth-masses. Analogs of our system seem scarce indeed.

The system around Kepler 20 is a good example of the complex problems encountered. It appears that there are two Earth-sized planets, 20e and 20f are between 1.7 and 3 times the mass of the Earth, if they are indeed rocky. Their diameters are 0.87 and 1.03 times that of the Earth and they orbit a Sun-like star with periods of 6.1 and 19.6 days, respectively. But to add to the complexity, three super Earths or Neptune analogs are present, sandwiched in between them, presumably another consequence of migration. So the sequence going out from the star is big, little, big, little and big. All are orbiting closer than our Mercury. The Kepler 20 system is a good example of the chaos surrounding the formation of planets. Any outcome seems possible to add to the problems of theorists.

The six planets around the red dwarf Gliese 581 demonstrate again more complexity, although these are not in coplanar orbits. One, Gliese 581d, has a mass 5.6 times that of the Earth (a super Earth or perhaps an ocean planet?) and is in the habitable zone. It appears to have a mainly CO_2 atmosphere (7 bars). Another, the infamous 581g, a rocky planet of about 4 Earth-masses, was claimed to have a circular orbit of 37 days in the habitable zone. Hence it aroused much interest and was even claimed to be the source of signals (implying intelligent inhabitants?), but eventually the "planet" turned out to be an artifact of the data.

Several other stars have well-packed planetary systems orbiting within a few AU, where the planets must have migrated inwards. Although many systems with multiple planets have been found so far there is nothing that closely resembles our own system. This is yet another illustration of just how unusual our inner planetary system is. These discoveries continue to illustrate the different outcomes of building planets.

How stable are such systems? The system around HD 10180 appears to have survived for about 4 billion years so it seems to be stable. It forms yet another example of how varied planetary systems can become. Clearly, lots of chance events occur during the formation of planetary systems.

Movements among the giant planets have dominated the architecture of our own system, hundreds of millions of years after the nebula had dissipated. Thus our apparently tidy system has arisen through the operation of essentially random processes. Systems that have giant planets in highly eccentric orbits are likely to have cleaned out both the inner and outer portions of the disk. Those in calmer environments seem more likely to have surviving rocky planets in stable orbits.

BODE'S RULE FOR EXOPLANETS?

> "We do not expect any universal rule on planet ordering to exist" [16].

Bode's Rule is a good example of the well-known psychological tendency to see patterns, something responsible for the "discovery" of so many false correlations in the natural sciences. The planets in the multiple planetary systems seen so far often seem to be spaced in a manner that recalls the Bode's Rule spacing of our planets. These arrangements are probably due to their mutual gravitational interactions, without significance for their formation. The regularities observed in these exoplanetary systems merely seem to show that migration is common and the resulting stable planetary systems tend to be spaced out in a regular fashion. Thus the final outcome of a multiple planet system is dependent on how many planets form, what size they are and how they interact due to their gravity. Stochastic events predominate.

Planets around binary stars

Most stars are present as binary systems so that planets are expected to have formed around either or both members. But searchers have tended to concentrate on single stars for obvious reasons. Binaries in wide orbits indeed harbor stars that presumably formed independently in the disks surrounding each individual star. Most planets seem to be in orbit around one member, although examples are known in which the planet orbits both stars. A good example is Tatooine (Kepler 16b) in which a slightly denser Saturn-mass planet is in a wide 229-day nearly circular orbit around a pair of stars. The planet was detected in transits around both of them, as the stars (0.7 and 0.2 solar-mass) are close together in a 41-day orbit and the stars and planets all lie in one plane.

Giant planets

These planets are the easiest to find, so that selection effects at present provide us with numerous examples. But giants are less common than smaller relatives as the sizes of planets peak towards low values (Figure 11). This effect is seen, for example, in the large number of small planets found by the Kepler mission. Neptune analogs seem particularly abundant and are being increasingly discovered. But among the giants themselves, many unexpected planets appear. Some are larger than Jupiter, ranging beyond the deuterium-burning limit, at around 12 or 13 Jupiter-masses, although the numbers fall off steeply beyond about eight.

Some have very large cores (e.g. HD 149026b), others have small cores (e.g. HD 209458b). These bodies extend far beyond what we are familiar with in our own planetary system. Radii for planets of similar mass may vary by a factor of two due to variations in temperature. Many have radii that indicate high density; others are grossly inflated [17]. Some are referred to as "oddballs", a comment that reflects how much we regard our own planets as the norm.

Because they have been the first of the exoplanets to be discovered, giants like Jupiter have provided us with much of what we know. Several outstanding features are present. The first is that there is a bimodal distribution of orbits. Many Hot Jupiters orbit in less than 10-day orbits. Mostly their orbits are eccentric, some reaching extreme values of more than 0.9, but peaking close to values of around 0.2. Clearly nature favors eccentric orbits and the near-circular orbits of our system are rare. There is a scarcity of bodies with masses greater than 10 or 12 Jupiter-masses orbiting stars. This used to be known as the famous "Brown Dwarf Desert" which extended up to 75 or 80 Jupiter-masses. Although brown dwarfs are true stars, sometimes with masses down to 3 Jupiter-masses, they mostly wander alone or in pairs, kicked out of the planet-forming nurseries by their larger siblings. They rarely orbit stars. This observation led to the concept of the desert. But nature continues to confound us. As noted above, a few bodies have been discovered in close orbits that have masses much greater than that of Jupiter.

Finally, some resolution of the debate over the origin of the giants can be discerned. Their properties support the core accretion model for their formation. Few data are consistent with the disk instability model, although the presence of giants far (100 AU) from their star is difficult to account for in any model. At that distance, forming planets by core accretion seems unlikely, as the density of the disk is too low. However, a giant formed by core accretion might migrate or be scattered outwards, so movement out to great distances remains as a possible explanation.

Free-floating planets?

Among the most unexpected discoveries by the gravitational microlensing method, is the apparent existence of a multitude of free-floating objects with planetary masses. Although only a small region of the sky was surveyed, extrapolation suggests that they are about twice as abundant as main-sequence stars. These bodies, sometimes termed "rogue" or "orphan" planets, might number 400 billion, although this assumes the presence of many smaller bodies. They are mostly about the mass of Jupiter and are at least 10 AU from any star. It is not yet clear whether these objects are truly floating freely in the void or bound to distant stars.

Are these creatures tiny brown dwarfs or planets? This raises again the question of what constitutes a planet? Thus they may be tiny stars, not planets. If they are planets, they were formed like other planets in disks around stars and were discarded and scattered out into space by gravitational reactions with larger planets.

Their huge number then suggests that perhaps planetary formation in disks around stars is ubiquitous and leads to much scattering of the products. We recall that forming planets is an inherently messy business now revealed as perhaps even messier than anyone thought.

These lonely wanderers (albatrosses of the universe) seem unlikely to generate much more human interest than the pulsar planets. They seem likely to become yet another example of nature's curiosities. The prospect of life on these bodies, which are not in habitable zones, warmed by a friendly sun, seems remote, although it is not beyond the imagination of theorists. Neither are they abundant enough to solve the problem of the missing dark matter.

The very abundance of free-floating objects, if they are cast-out planets, demonstrates the dominant role of chaos in making planets. So they cast a somber shadow over the chances of finding Earth-like planets sitting comfortably in habitable zones for periods of several billion years.

NEPTUNE-MASS PLANETS

These are very common, several times more abundant than Jupiter-mass planets. Although giant planets seem more common around more massive stars, Neptune-sized planets seem more abundant around low-mass stars. But these correlations may be due to selection effects.

This abundance of many Neptune-sized planets presents interesting problems for the formation of planets. Why are they so common? Are they mostly composed of rock covered by a low-mass gas envelope? Or are they formed of lower-density ices with a thin sheath of gas, or perhaps some mix between? The few that have been observed in transit have densities indicating that they have large (80–90%) cores of ices and rock covered by a small gas envelope, just as in their namesake.

Is this low gas content an original feature or was it caused by evaporation of a larger gas envelope? That is, are they remnants of former giants like Jupiter? Such an origin was often proposed to explain our Uranus and Neptune. There is some evidence that suggests otherwise. Those Hot Jupiters which orbit within a few million miles from their star at temperatures around 2000 K, surprisingly do not lose much mass. This tells us that it is difficult to get rid of gases from giant planets by evaporation. The notion that the low gas content of Neptune-mass planets is due to evaporation of a former gas giant would thus seem difficult to achieve.

The abundance of Neptune-mass planets indicates that Nature apparently has little difficulty in forming cores with many Earth-masses. This seems in accordance with the notion that many ice and rock cores form at the snow line. But they must form fast enough to capture some gas before it is lost. This is a general explanation for Uranus and Neptune in accordance with the scenario predicted by the core-accretion model for building planets. In this model, cores of 10 Earth-mass or more form near the snow line from the accumulated ices and dust. But how much gas they manage to capture is a chancy affair, depending both on the time it takes for the core to assemble and how long the gas lasts in the nebula.

The disk instability model predicts no such correlation. This model predicts that giant planets should reflect the overall composition of the disk, except that gas envelopes might be removed by subsequent evaporation. But as noted above, this seems difficult. This seems unlikely to be a common process because Neptune-mass planets are very abundant.

Super Earths

Those planets between Earth-mass and about 10 or 15 Earth-masses are often classed as "Super Earths". They are interesting exoplanets as we have no examples of such in our planetary system. In the absence of much information about them, they might be smaller editions of Neptune or larger specimens of the Earth. Whether they will bear any resemblance to our own planet is a moot point. No doubt science fiction writers will soon be populating them with creatures able to withstand the crushing gravity on their surface.

They have presented students of exoplanets with perplexing problems. Unknown here, they are extraordinarily common, forming perhaps 30% of exoplanets.

Current models suggest that only planets that have more than about 6 Earth-masses can retain an envelope of gas that can amount to as much as an Earth-mass. In such a case, they might be mini-Neptunes with thick atmospheres. But this illustrates once again, the hazards of attaching evocative labels. Where does the boundary lie between rocky planets and ice giants? Is there one? Probably there is a continuum and both triangles in Figure 9 will eventually be filled.

To make matters worse, super Earths or mini-Neptunes in orbits close to their stars seem to be very common. Current models for the migration of planets usually involve interaction with the gaseous disk of the nebula. This is the classic explanation for Hot Jupiters. But rocky planets form long after the gas has gone, and this presents a current problem for their formation, in terms of the Standard Model, if these bodies are really scaled-up versions of the Earth or Venus. If they formed after gas had left the inner disk, their migration inwards is presumably due to gravitational interactions with other planetary bodies.

The difficulties of dealing with these potential super Earths are well illustrated by the 4.8 Earth-mass planet, Corot 7b. This planet is locked in synchronous rotation, having an orbital period of less than one day. Dayside temperatures are about 2000 K. If it has an Earth-like composition, it would have an iron core 5100 km in radius, under a rocky mantle 6000 km thick with a thin upper mantle about 300 km thick. In contrast, the upper mantle of the Earth is 600 km thick. The planet would generate several times as much heat as the Earth, if it had a similar component of radiogenic elements. The surface on the sunlit side might well be molten or resemble a sort of super-Io. It is not clear whether it was originally an ice giant like Neptune that had lost its ices, or whether it had always been rocky.

Another example is Kepler 10b, already mentioned, one of the smallest to date and an undoubted "rocky" planet. Again its orbital period is less than 1 day, so the planet is presumably locked into a synchronous orbit [18]. A much-heralded example is Kepler 22b, 600 light years away in a "habitable zone" with an orbital period of 290 days around a Sun-like star. It is 2.4 times the radius of the Earth or about half that of Neptune and is somewhere between 10 and 35 Earth-masses. Whether it is gaseous like Neptune or rocky and is "the best hope for human habitation" remains unknown. Whether it has "continents, oceans and creatures already living on its surface", as one news release stated, remains to be seen. Attempts to discover lower-mass planets are currently limited by stellar jitter.

Complementing the Kepler mission in space are Earth-based searches that can employ very high-resolution spectrometers, such as the ESO HARPS instrument at La Silla, Chile. This has discovered a 3.6 Earth-mass, possibly rocky planet in a habitable zone around its parent star (HD 85512). This star is much cooler than the Sun and the orbit of the planet is only one quarter of the Sun–Mercury distance.

Although liquid water might exist on its surface, the planet may again be locked into a synchronous orbit.

A large literature has developed, attempting to predict the internal composition of these bodies. But as most of these models are simply scaled-up Earths, little is learned from these exhaustive efforts of extrapolation. As one author laments, "In reality, deducing exoplanet interior composition is very difficult" [19].

The present state of the Earth, including its interior, oceans, continents and atmospheres is the end result of a long sequence of stochastic events as described in Chapter 6. The nature of our mantle and core still remain fertile fields for study with much controversy. It is worth recalling that the Earth is an unusual planet even by the standards of the solar system. Venus, our twin planet, glows brightly in the evening or morning skies to remind us how differently planets of similar mass and composition may evolve. So extrapolating from the Earth to predict what the interiors of super Earths might look like seems hazardous.

But the presence or location of super Earths or mini-Neptunes should not be surprising. Planets, in contrast with stars, are formed by chaotic processes so that many outcomes are likely. Our solar system becomes even more of a rarity.

EARTH-LIKE, ROCKY OR TERRESTRIAL PLANETS

> "Terrestrial planets are of major scientific importance because they are the planets suitable for life as we know it" [20].

It is very difficult to get a precise definition of what exactly is meant by "Earth-like" planets as distinct from those with "Earth-mass". Planets with less than about 15 Earth-masses are commonly called terrestrial, although there seems no particularly good reason for the term. Even the four rocky planets in our system show little similarity and could just as well belong in separate systems. The Earth is unique even within our solar system and so forms a poor yardstick for comparison with exoplanets. Presumably, what workers seem to have in mind is something close to the Earth in mass and radius. Some optimistic estimates suggest that as many as one third of Sun-like stars have a rocky planet in orbit within a habitable zone. But even if such planets have liquid water on their surfaces, this is unlikely to be a sufficient condition for intelligent life to develop.

The standard model for forming our rocky planets is to assemble them out of the rubble left over in the inner nebula. This was depleted in gases, ices and the more volatile components of the rock fraction. It was from this dry rocky rubble that our terrestrial planets formed. Is this the case elsewhere? In the absence of any direct evidence, it seems necessary to assume so.

Classifiers have proposed many different sorts of terrestrial planets. These include those such as "Mercury-type" where iron dominates, or silicate planets that might resemble our Moon. But such endeavors at labels for planets are as doomed as attempts to classify the satellites in our system or to place the planets themselves into neat pigeonholes. Planets in circum-stellar disks are formed by stochastic processes and so tend to be unique.

The system around Kepler 20 has already been discussed. Although it contains two planets close in size to the Earth (interspersed with larger companions), one needs to recall the small, apparently crucial size differences that exist between the Earth and Venus.

The question of the likely number of Earth-mass planets remains unresolved at present although it is expected to be large. While mass, size and orbits can be established now, further progress will depend on dedicated telescopes in orbit that can take spectra of the atmospheres of likely candidates. Perhaps new, less expensive techniques will become available. But the interest surrounding this particular facet of exoplanets serves once again to illustrate that the search is mostly for something like us out there.

MOONS

Moons are expected to be common around giant planets, from analogy with our system. They might either form from a sub-nebula around the giant or merely be captured debris and some might even be habitable. But because none of our 160 or so satellites resemble one another, look-alikes for Triton, Titania or Titan are not to be expected elsewhere. Neither would the discovery of such satellites of giants do much more than satisfy our curiosity, unless they were habitable.

The real interest is whether a Moon comparable to our own might be in orbit around an Earth-mass planet that finds itself in the habitable zone. Such a satellite might stabilize the tilt of the planet, making it a more suitable place in which life could develop. But there are many assumptions, including whether the planet even has a tilt or whether a stabilizing moon is really necessary.

The process that formed the metal-poor Moon was probably responsible for both the obliquity of our planet and for its 24-hour rotation period, both properties affecting biological evolution. As Jacques Lascar and co-workers have noted, without the presence of the Moon, the obliquity of the Earth would be "chaotic with large variations reaching more than 50° in a few million years and even, in the long term, more than 85° in the absence of the Moon" and so "our satellite is a climate regulator for the Earth" [21]. But whether this effect is unique to the Earth remains unknown. Another effect of the giant Moon-forming collision was to remove any primitive atmosphere from the Earth. But whether moons are necessary or common around Earth-mass planets remains to be discovered.

The collision that formed our Moon depended on special conditions such as the angle of impact and the velocity of the impactor. Of course, other orbiting satellites might stabilize a planetary tilt without requiring such special parameters. But this assumes that the planet needs an agreeable tilt that ameliorates climate extremes.

Several modelers have considered the problem of forming moons around rocky planets. All depend on many assumptions and approximations, some concluding that our Moon is a "cosmic rarity". Others suggest that 10% of Earth-mass planets in habitable zones could possess moons that stabilize their tilt, whatever value it may have. Perhaps some planets like our Venus are not tilted appreciably. While one must admire the prodigious intellectual and computing power employed in such calculations, possibly, it will be better to wait for observations in view of the huge uncertainties involved in trying to make predictions about an essentially random process [22].

WATER WORLDS OR OCEAN PLANETS

These are generally supposed to be rocky planets with deep oceans. They form yet another speculative variety, sometimes perceived as Uranus- or Neptune-like bodies without any gas envelope. In this manner, they might be large-scale analogs of bodies like Ganymede, Europa or Callisto. They are expected to have deep oceans where real estate above water level is scarce or non-existent.

Perhaps conditions in an early nebula might favor the formation of water-rich planets. Most theorists seem to imagine an Earth submerged beneath a deep ocean. But this requires a delicate balance. To form such a body, ices such as methane and ammonia, but not water, need to be lost from the inner nebula or hidden in the core. In our examples of Uranus and Neptune, we see giants with large amounts of the other ices. But if the temperatures are right, methane and ammonia might be lost, leaving water as the principal ice. This seems to be the case in the region of Jupiter and Saturn, where the satellites appear to be dominated by water ice. So water worlds such as super Ganymedes, Europas or Callistos may exist, but would require some special conditions for their formation. But theorists have little difficulty in forming water worlds with over 100 oceans of water in simulations past the snow line [23].

CARBIDE PLANETS

Among other questions worth considering is whether rocky planets are necessarily formed from silicates which is the norm here. As discussed above, it has been suggested that planets forming around stars with high C/O ratios might form carbide rather than silicate planets [24].

But this supposes that the planets will mirror the composition of the star. Planets indeed form preferentially around metal-rich stars, but none of our planets has the composition of the Sun. Either they are enriched in metals, as are the giant planets, or depleted in volatile elements, like our rocky planets. Carbon is mainly present as methane (CH_4), CO or CO_2 ices in the primitive nebula. These ices are swept away from the inner disk before the rocky planets form. So no matter how high the carbon content of the parent star or the primitive disk from which it forms, the ices that contain most of the carbon seem unlikely to survive the strong winds emanating from the star in its early evolution. The inner parts of the disk where rocky planets are going to form will lose much of their carbon, so making the formation of carbide rather than silicate planets less likely.

PLATE TECTONICS

The fact that "continental drift" was not generally accepted in the Earth sciences early last century has become a cliché. This example has been commonly used to defend bizarre new theories about the origin or evolution of the Earth, or of features such as impact craters. But science does not deal with beliefs based on controversial interpretations. This is the realm of mythology. Once the movement of the plates on the surface of the Earth was established by solid geophysical evidence, the concept was accepted rapidly as a consensus view, in accord with the nature of science.

The possibility of the development of plate tectonics on exoplanets is often given as a requirement for habitability in models of Earth-mass planets. On Earth, the operation of planet tectonics is responsible for building continents. These formed the necessary platforms for the development of intelligent life, although curiously, *Homo sapiens* appeared on only one plate among several possible stable environments. But it remains a moot question whether the development of plate tectonics on exoplanets is either inevitable or a requirement for habitability of planets.

There are two end members for the evolution of the mantles of rocky planets usually referred to as the "stagnant lid regime" and the "plate tectonics regime". In the first case, the surface of a rocky planet forms one plate. Examples in our own system include Venus, Mars, Mercury and the Moon.

Venus presents us with an especially interesting case. The planet was resurfaced with a basaltic crust about 750 million years ago. The previous 4 billion years of Venusian history have either been erased or covered over by extensive volcanism. Among the many models that have been proposed, episodic periods of lava extrusion and foundering of a former crust seem most likely. Venus has about the same heat production as the Earth, but models suggest that it is released every billion years or so in a splurge of volcanic activity. The surface geology of Venus indicates that such episodes might last a few million years before the mantle cools down and begins slowly heating for the next cycle.

There is considerable disagreement among geophysicists about whether plate tectonics is likely to develop on planets larger than the Earth, with completely opposite conclusions being reached in current papers [25]. One attempt to resolve this issue comments that "scaling laws predict that the more massive the planet, the more likely the occurrence of plate tectonics" [19], but this extensive study concludes that "scaling of the driving to resistance forces is not yet available to determine whether planets more massive than Earth are likely to have plate tectonics" [19, p. 390].

Whether we will ever have enough geophysical data to resolve this significant issue seems doubtful, much like the question of moons around Earth-mass planets. Such calculations based on geophysics rely heavily on properties such as temperature, pressure, viscosity, composition, density, size, heat capacity and bulk modulus, all used to calculate "equations of state" but all subject to considerable uncertainties. Assumptions and simplifications abound. Although some theorists find that plate tectonics will develop on super Earths, others dispute this finding [25].

Geophysical models that discuss the possibility of plate tectonics on planets more massive than the Earth, concentrate on ways of breaking the surface plate, after which subduction of the surface is expected to begin. But examination of plate tectonics on our planet reveals a more complex picture. Our plates are dragged into the mantle by the subsiding slab that has transformed into denser phases by increasing pressure at depth. This pulls the slab down so that pull dominates push by a factor of ten. Our plates slide on the asthenosphere, a zone about 100 km thick of partial melt lubricated by water, that occurs around 100 km beneath our oceans and continents.

The mantle above the down-going slab likewise produces melts due to an influx of volatile elements and water, driven off the slab by increasing heat as the dense slab sinks below 100 km or so. This results in the eruption of andesitic volcanoes, like Mt. St. Helens, which provide the raw material for making continents. Water contents of a few hundred ppm (the Earth has about 500 ppm) are therefore crucial. As the aphorism states, "No water, no granites; no oceans, no continents" [26]. Breaking of plates due to thermal stresses may indeed occur on a dry planet [25], but seems unlikely to lead to any scenario resembling our habitable planet.

On Earth, only the oceanic crust is recycled back into the mantle. Continents are too low in density and too thick to be dragged down into the mantle. Probably water content, geophysical parameters and tectonic processes are unique to each rocky planet, as happens in our system.

But speculation continues nonetheless. In a discussion of the possibility of the occurrence of "Earth-like" planets around 47 Ursa Majoris, a star that has two gas giant planets orbiting at 2.1 and 3.7 AU, the authors state that "we assume an Earth-like planet with plate tectonics, a crucial ingredient for our models". They concluded by predicting the existence of such a body [27]. This indeed seems to be building the answer into the question and so making their calculations self-fulfilling.

One might have thought that the absence of plate tectonics on our twin, Venus, would have given pause to such speculations. The Earth is a unique planet within our system with a unique water budget, the consequence of stochastic processes that resulted in the development of plate tectonics. Logic suggests that our planet is not the best example to scale from.

CLONES OF THE EARTH

"We are good at looking for things like ourselves" [28].

There is a surprisingly common perception that Earth-like planets will be found in extra-solar planetary systems, although exactly what is meant by "Earth-like", "Earth-type", "Earth-mass", "terrestrial" or "rocky" planets is rarely made clear. Although the Earth is a unique planet in our system, the notion that the Earth represents a norm from which others differ to varying degrees seems deeply entrenched.

So how likely is the appearance of a clone of the Earth? Indeed no general principles that might be applicable to "Earth-like" exoplanets and the question of their habitability appear from this investigation. The tale told of the development of planets in our system reiterates that obtained from a study of the exoplanets: random processes dominate.

Stars differ mostly in mass, and form dominantly from hydrogen and helium with a dash (from zero to a few percent) of metals (ices plus rock). But planets may form from any combination of gases, ices and rock (Figure 9). But even within this system of three components, there are complexities. The respective abundances of the ices (water, methane and ammonia are the most common) may vary, while the rock component may be that of the nebula or variously changed by processes, such as collisions, or through the loss of the more volatile elements, within the nebula [9]. Information from our own system reveals the obvious requirements for the formation of planets of metals, orbits of low eccentricity and the avoidance of giant planet migration into the inner nebula, which might have cannibalized the terrestrial planets [8].

The inner nebula, from which the rocky planets formed, was also depleted in elements whose sole common property is volatility. Ironically, this depletion, which occurred in the habitable zone around the Sun, included a large fraction of the biologically important elements, such as C, N, P and K. Of course life elsewhere need not conform to life here. Silicon-based life is a favorite of science fiction writers, but not of biologists or chemists.

Collisional events may change the metal-to-silicate ratios on planets, Mercury being an outstanding example. Although often thought to cause further loss of volatile components, it is even difficult to lose elements as volatile as sulfur from small bodies like the Galilean satellite, Io. Even the gases seem reluctant to leave Hot Jupiters. If the high density of Mercury is indeed due to a collision, as argued later, it has not resulted in the loss of elements as volatile as potassium.

The inner nebula was also bone-dry, as shown amongst other evidence by the anhydrous nature of the primary minerals (olivine, pyroxene and plagioclase) of meteorites, so that the splash of water for our oceans had to come in later from the neighborhood of Jupiter. If Earth-mass planets in habitable zones form like the Earth, the presence of water and so plate tectonics must be expected to be a random variable. The presence or absence of a moon might also be crucial, as noted above.

When nature got around to building two similar planets, it finished up with the Earth and Venus. These twins, unlike the earlier-formed Mars and Mercury, are close in mass, density, bulk composition and in the abundances of the heat-producing elements (potassium, uranium and thorium), but the geological histories of these "twin" planets have been wildly different and Venus provides no haven for life.

Indeed, a sober contemplation of our own solar system, the only example available for close study, reveals just how difficult it is to make planets that resemble our own. None of our planets or their 160-odd satellites resemble one another, except in the broadest outlines.

Our own eight vary from gas-dominated (Jupiter), though ice and rock mixes (Uranus and Neptune) to rock and metal mixes among the terrestrial planets. Metal may dominate, as in Mercury, or rock if we include the Moon. But even the composition of the rock component may differ, not only from that of the original nebula, but also between the Moon, Earth and Mars, which are successively richer in volatile elements. The other 160 assorted satellites likewise fail to supply us with clones. None of this diversity encourages us to derive general principles for forming crusts. The results of all these factors are beginning to appear among the exo-planetary systems. In summary, the difficulty of producing clones of our present solar system makes its duplication unlikely.

5 Our solar system

> "Saturn's rings were left unfinished to show us how the world was made" [1].

How does our solar system fit among the great host of exoplanetary systems? It is not a typical sample of what is out there so it is not a good example of the "Principle of Mediocrity". Its obvious advantage is that it is the only one that we can examine in detail and so forms the basis for current models of planetary origin. What is revealed are not only our eight planets, interesting though they are as possible analogs for exoplanets, but also hosts of other bodies in orbit either around the Sun or planets. We have over 160 satellites, four ring systems, dust, swarms of many different species of asteroids, KBOs, TNOs and wanderers such as comets, centaurs, NEAs and NEOs. The system resembles a half-finished long-abandoned building site, littered with leftover rubble and assorted debris. Although little of this detail is yet evident in exoplanetary systems, it is useful to examine our solar system in depth to see what actually happened in one disk around a star. It is not only a matter of just forming planets.

The solar system has long been a benchmark, becoming from familiarity, what was commonly expected to be a norm. But reality was different, as the first exoplanet discovery, Pegasi 51b, demonstrated so dramatically. Now even our system of planets is seen to be uncommon, the end result of chaotic and chance events.

There are so many excellent descriptions of our system available [2] that it is pointless merely to repeat well-known data here. Instead I discuss here some salient features of the solar system that are important to our understanding of exoplanets. Although this includes such exotica as planetary rings and centaurs, the discussion focuses on planets, with particular attention to the rocky terrestrial planets. Although we have only a subset of possible planets, we can expect that eventually examples will be found filling not only every possible composition between rock, ice and gas [Figure 9] but also a wide range of masses and eccentricities [Figures 11, 12].

OUR STRANGE PLANETARY SYSTEM

Books about our solar system used to follow a traditional path, starting with Mercury and marching stolidly outwards, wending their way past the other rocky planets, trawling through the asteroid belt, marveling at the giants and finishing up at the strange dwarf, Pluto. This picture became established as the very model of a modern system of planetary arrangement and indeed of what was to be expected elsewhere. The Solar System in fact is not very tidy and the arrangements of the planets are not what might be expected from simple condensation from the original disk of gas and dust. Neither, for that matter, does it look like the work of a competent designer with omnipotent powers. What is the reason for this diversity?

What has also become apparent is that the very architecture of the system with its four inner rocky planets, the asteroid belt, the four stately giants, the Kuiper Belt and the Oort Cloud of icy planetesimals, far from arising from uniform processes, has arisen through accidents, chance events and migrations among the giants. Such random events are even responsible for such details as the pockmarked face of the Moon.

When I wrote *Destiny or Chance*, in 1998, I tried to explain how our strange variety of planets and the rest had come about. Even then it was obvious that Pluto was a misfit among the planets, to the dismay of school children and sentimentalists alike. It was clear enough that the multitude of components that constitute the solar system had arisen through a series of chance events, as unlikely to be repeated elsewhere as winning the lottery.

Even among our familiar rocky planets, Mercury with its extreme density is an anomaly, Mars is tiny while our twin planet, Venus, is bizarre by earthly standards. Then there is a gap with no planet but occupied by swarms of rubble. This is the asteroid belt, whose nearer rocky members have been melted, but those further out contain ice, have a different source and are related to comets. The giants, Jupiter, Saturn, Uranus and Neptune are all so different from one another that they could be inhabitants of different systems without exciting comment. Next there is the abrupt gap beyond Neptune, followed by the zonal arrangement of the Kuiper Belt with its several populations of tiny icy bodies. Finally the Oort Cloud, as though to complete the picture, forms a kind of halo to the whole.

How did this arrangement, so long familiar that it became to be accepted as the expected norm for planetary systems, come to pass? Now we have enough information about exoplanets that we can attempt a beginning at answering the uniqueness question. A solution will only appear when we have a complete inventory of planetary systems, but the prognosis for something resembling ours looks bad at present.

A brief current history

Following the formation of the Sun, the inner nebula was swept clear of gas, ices and the more volatile elements of the dust fraction. Much of this material piled up at the snow line [Figure 8], a location at which water ice could condense. This was somewhere between 3 and 5 AU. Sunwards of the snow line, only dry gas-poor rubble was left, from which the four rocky planets would later form. The increased amount of material at the snow line enabled the rapid growth of four large bodies of over 10 Earth-masses. These formed the cores of the giant planets. Current models suggest that the core of Jupiter formed around 3.5 AU, that of Saturn a little farther out, near 5 AU, while those of Uranus and Neptune formed at 6 and 8 AU respectively. These events occurred within 1 million years of the formation of the Sun. Gas was still present and was captured in decreasing amounts by all four cores. Satellites formed from the mini-disks surrounding the giants. Various wanderers were captured then or later by the giants, forming

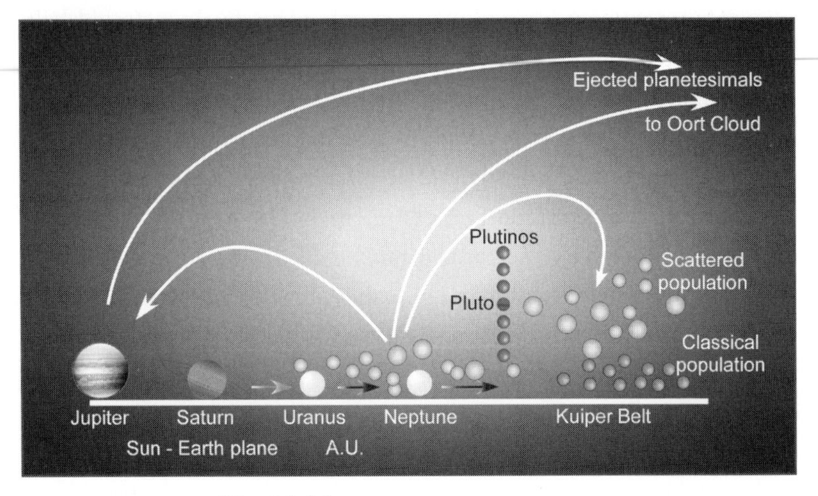

FIGURE 14 Nice Model

A representation, not to scale, of one stage of the migration of the giant planets according to the Nice Model. Saturn, Uranus and Neptune are moving outwards, scattering the residual icy planetesimals out into the Kuiper Belt and the Oort Cloud. Some, including Pluto and the Plutinos have been captured into resonant orbits with Neptune. (Adapted from Morbidelli, A. and Levison, H. *Nature* Vol. **422**, p. 31, 2003.) See also color plates section.

the irregular satellites. The four giants began migrating while gas was still around.

Current models again suggest that Jupiter and Saturn both moved inwards, depleting the regions of the asteroid belt and Mars before reversing and pushing Uranus and Neptune out. Beyond lay a mass of icy planetesimals extending to about 35 AU which seems to have been the original limit of the solar nebula.

Gravitational interactions between these bodies and the giants caused the monsters to move slowly out for the next few hundred million years, scattering the smaller bodies both outwards and inwards. Uranus and Neptune were pushed further out. Here they rammed into the disk of planetesimals, scattering them both outwards to form the Kuiper Belt and Oort Cloud (Figure 14) and inwards to form the outer asteroids and Kuiper Belt comets.

Meanwhile, much had occurred sunwards of Jupiter. The asteroid belt was populated by survivors of the rocky inner planetesimals, supplemented by arrivals of ice-rich bodies from the outer solar system.

The small, dry, rocky planetesimals left there grew within a million years to Moon-sized bodies and quickly to the size of Mercury and Mars. Many melted and differentiated from short-lived radioactive heating or from the energy of collisions. Then began a slow process of accumulation of these bodies of all sizes to form the Earth and Venus, taking perhaps a 100 million years to complete the series of collisions. Among the last of these, a Mars-sized body plowed into the Earth, producing the Moon.

When did the solar system form? Should we set zero time as the time of separation of the molecular cloud, the formation of the disk around the Sun, or that of the Sun itself? None of these events can be measured precisely. The most convenient marker for the beginning of the solar system is the age of the oldest solid bodies found in meteorites. The oldest most reliably dated objects are the inclusions of refractory minerals. These are centimeter-sized grains of high temperature minerals that have been through repeated episodes of condensation and evaporation in the turbulent disk near the Sun. These grains formed 4567 million years ago, a date easily remembered and precise to within a couple of million years. Although this event (T_{zero}) so long ago is incomprehensible on human timescales, the universe had already been around for three times as long.

THE ARCHITECTURE OF OUR SOLAR SYSTEM

Our system, familiar on every schoolroom wall, appears as fixed and regular as the Earth itself. At a glance, it looks simple enough, with four small rocky planets, an asteroid belt, four giants all in tidy near circular orbits, followed by a Kuiper Belt of small icy bodies with an Oort Cloud beyond. Looked at objectively, it is a curious arrangement, but its very familiarity has obscured its problems. This apparent regular arrangement has long been an intriguing topic. Most famously, that the planets are spaced out in a kind of geometrical progression, celebrated as Bode's Rule, which has sometimes achieved the status of a law.

Bode's Rule [3]

This curious mathematical relationship has fascinated students of the solar system for over 200 years. Much intellectual energy has been expended in trying to account for it and the "rule" was commonly cited as one of the significant features to be satisfied in any theory for the origin of the solar system. It led us to expect that other planetary systems should follow a similar spacing and so it was sometimes called Bode's Law.

This interesting relationship was first discovered by Johann Daniel Titius von Wittenberg (1729–1796). It was later drawn to popular attention by Johann Elert Bode (1747–1826) and so is correctly termed the Titius–Bode Rule.

The distances of the planets from the Sun can be expressed as a series: 0.4, 0.7, 1.0, 1.6, 2.8, 5.2 etc. that is close to their separation in AU. This sequence of numbers can be arrived at by adding a constant 0.4 to the sequence of 0, 0.3, 0.6, 1.2, 2.4, 4.8 etc., although many more sophisticated formulae were developed for what is a geometrical progression. When the asteroid Ceres was discovered at the location of the "missing planet", at 2.8 AU in between Mars (at 1.6 AU) and Jupiter (at 5.2), the "rule" achieved the status of a cult. But the relationship is only approximate. It works well enough out to Uranus, but breaks down at Neptune. This planet should be out at 38.8 on Bode's scale, but it resides inconveniently much closer in at 30 AU.

Does the Titius–Bode Rule have any real significance? It seems reasonable to expect that if the rule had some major physical significance in planetary systems, then some other properties might also vary with the simple mathematical regularities in spacing of the planets. However, there is no correlation in our system with mass or composition, either with the spacing given by the rule, or with distance from the Sun. This raises the possibility that the rule is a secondary, rather than a primary, property of the solar system.

The twin concepts of the solar nebula and Bode's Rule led to the notion that the planets had formed in place, perhaps condensing from rings in the nebula as Laplace had imagined. But a fatal objection to the significance of the "Rule" is that we now understand that the present spacings between the planets have arisen naturally by tidal forces. Much more recently it has been realized that planets migrate, so the famous rule is not some kind of universal rule of thumb for making planetary systems, and accordingly has become less significant except to students of the dynamics of our early solar system. It has become a matter of inter-planetary dynamics.

THE NICE MODEL [4]

Fast computers have replaced the tedious numerical calculations that beset Titius, Bode and a host of other worthies. An elegant model, named for the pleasant town in France where it originated, has now arisen to suggest how much of the current pattern may have arisen and its success has generated wide support.

The model is briefly described here. Jupiter formed at about 3.5 AU. The cores of Saturn, Neptune and Uranus in that order, formed near the snow line, but further out at 4.5, 6 and 8 AU, where the gas content of the nebula was diminishing. Beyond Uranus lay the remnants of the nebula, a cloud of thousands of icy planetesimals (about 35 Earth-masses) that extended out to about 35 AU, a little past the present orbit of Neptune.

Encounters of these planetesimals with the giant outer planet, Uranus, scattered their orbits. Originally they were scattered inwards and the planets moved out a little, to conserve angular momentum. As the small icy bodies moved inwards, so successively encountering Neptune and Saturn, these giants in turn shifted outwards. These gravitational changes were minute but slowly accumulated. Finally the planetesimals interacted with Jupiter. Messing with giants is never recommended as many a fairy tale attests. The mighty gravitational field of Jupiter scattered these tiny interlopers far and wide, some into elliptical orbits and others out of the solar system.

These slow movements might have continued indefinitely but after several hundred million years into the model, Jupiter and Saturn moved into a 1:2 resonance. Their orbits became eccentric, creating havoc and destabilizing the whole system. Jupiter wielded its great gravitational influence, pushing Saturn out to its current location. Saturn in turn swept Neptune and Uranus into eccentric orbits. These ice giants drove out into the great disk of icy bodies, a process that destroyed much of the original solar nebula. Neptune and Uranus swapped orbits in the mayhem, but finally their orbits damped down into their currently near circular form through interacting with the residual planetesimals.

The model accounts for many of the observed features of the solar system. As Neptune moved out by perhaps 10 AU, it scattered the remaining icy bodies. Some, Pluto being the famous example, were captured into resonant orbits with Neptune, while the giant captured Triton. Other icy bodies were sent into highly inclined and eccentric orbits, forming the scattered population of the Kuiper Belt (Figure 14).

A standard criticism of the Nice Model is that it fails to account for the so-called classical population of the Kuiper Belt, those icy bodies in low inclinations. But they seem easily enough explained as residual from the original Laplacian disk, although moved out a bit, while the scattered population originated farther inwards near Jupiter. These two sets of KBOs are certainly distinct in composition. Other icy planetesimals were scattered out into highly elliptical orbits and now form the halo of the Oort Cloud.

The survivors of the population that were scattered inwards now form the Jupiter Trojan asteroids with their large eccentric and inclined orbits, the Hilda asteroids and the outer (D-type) asteroids of the Main Asteroid Belt. These latter contain a little ice and carbonaceous material. Even Ceres, the largest asteroid and first to be discovered, may be one of these interlopers. The "Trojan" asteroids found around Neptune belong as well to this period while the giants captured many bodies that now form their irregular satellites.

During all this mayhem, many collisions occurred that are preserved as craters on the satellites of the giant planets. The most dramatic local effect was the irruption of these bodies into the asteroid belt. This sent a shower of impactors sunwards that resulted in the Late Heavy Bombardment, whose effects are so dramatically evident on the Moon. The dating of this event, around 4 billion years ago, has long been a puzzle, no model previously being able to account for such a late unique event. The basic problem was where to store the impacting bodies in space for several hundred million years before they hit the Moon.

The principal attraction of the Nice Model is that it accounts for many of the curious features of our system, such as the Trojan asteroids, the irregular satellites of the giant planets and the Kuiper Belt. It suffers from the notion of being "untestable" in the same sense that leads to complaints that much of geological history or other events that occurred long ago cannot be reproduced. But the predictions from the model give a good match to what we observe so that it remains the best current explanation. Whatever the ultimate success of the Nice Model, the major conclusion remains that there has been much rearrangement of the planets since they were formed. This operation of such chaotic events casts another shadow over the search for a planet like the Earth, comfortably seated in its habitable zone, or for solar systems like our own.

When I wrote *Destiny or Chance*, in 1998, I was impressed by the extraordinary variety of planets, their satellites and all the other assorted individuals in our system. Nothing resembled anything else, an observation that led me to suppose that it would be very difficult to produce clones. But now the very architecture of the system itself is seen to be chaotic, subject to chance events.

The giant planets

> "Jupiter's orange and yellow bands are so roiled up that its disk might have been painted by van Gogh at Arles. Saturn, with more delicate bands of ochre, resembles a Monet haystack in a sunlit mist. Uranus' disk, though, is so featureless and limpid as to suggest the still pond around a Monet water lily" [5].

Although many giants have been found among the exoplanets, our four are the only examples currently available for close inspection. Accordingly, it is useful to see what they have to tell us about what we might expect further afield.

Jupiter and Saturn are splendid planets. With their associated rings and satellites, they excite the admiration of all observers. A consensus seems to have been reached, although still awaiting final proof from the Juno mission, that Jupiter indeed has a large rock and ice core (somewhere between 8 and 38 Earth-masses), while it is enriched in metals between two and six times compared to the Sun. Both features follow naturally from the core accretion model.

Current models suggest that the core of Jupiter formed near the snow line, possibly at around 3.5 AU in the middle of what is now the asteroid belt. Being the closest to the Sun, it managed to capture about 300 Earth-masses of hydrogen and helium gas. This massive planet then dominated subsequent evolution of the solar system. It seems to have moved inwards, perhaps as close as the current orbit of Mars, depleting the asteroid belt that now contains only 5% of the mass of the Moon. The region where Mars would later accumulate was likewise depleted, so that the presence of the giant is responsible for the small mass of that planet, Mars being only 1/3000 as massive as Jupiter. But under the influence of Saturn, the giant seems to have tacked (to use a nautical analogy) at about 1.5 AU and moved back out, allowing rocky remnants to populate the inner parts of the asteroid belt.

But much might have turned out differently. If the Sun had had a more violent early history, the gas might be swept away before a core became big enough to capture gas. A system with several ice giants or super Earths like Uranus and Neptune might have resulted. We now commonly find such systems. We might have had a larger version of Mars, or a planet in place of the asteroid belt and a much smaller, perhaps rocky or icy version of Jupiter. Clearly the timing of the formation of Jupiter relative to the sweep out of gas was critical. Jupiter might have migrated close in to the Sun, like the Hot Jupiters in other systems. Such migration into the inner nebula could destroy or prevent the formation of inner rocky planets in its march, just as the capture of Triton by Neptune probably demolished an earlier regular satellite system around that planet.

Saturn is famous among planets for its ring system which is familiar to every child. The Babylonian astrologers recorded Saturn in the seventh century BCE. Named after the Roman god responsible for agriculture, it was the outermost planet known until the discovery of Uranus in 1781, followed by that of Neptune in 1846. The presence of a large core in Saturn (between 13 and 28 Earth-masses) is long established, so that the planet is strongly enriched in metals between 6 and 14 times the composition of the Sun, consistent again with the core accretion model. Models suggest that this core seems to have formed around 4.5 AU at the snow line, following which it managed to collect about 80 Earth-masses of gas.

Uranus was the first planet to be discovered since antiquity, although its discovery by William Herschel (1738–1822) in 1781 was the accidental outcome of a star search. He called it George's Star (Georgium Sidus) after King George III, who then presented Herschel with a lifetime pension. Such nationalistic fervor, although rewarding for Herschel, did not receive universal acclaim. The planet was eventually given the appropriate classical name of Uranus.

However, it was soon discovered that the orbit of Uranus was being affected by another large body further away from the Sun. Study of these variations in the orbit of Uranus led eventually to the discovery, in 1846, of the apparent cause. This was the existence further away from the Sun of another large planet, Neptune.

The discovery of Neptune represented a triumph for the Newtonian System but it is not without irony and forms a cautionary tale. J. C. Adams (1819–1892) in England, and Urbain Le Verrier (1817–1877) in France, independently calculated where a planet might be found on the basis of its effect on the orbit of Uranus. By September 1845, Adams had worked out where to look, but there was bureaucratic delay in England in acting on his predictions. Meanwhile, Le Verrier was having similar difficulties with his French colleagues, and finally persuaded the astronomers at the Berlin Observatory to search. The planet was discovered by J. G. Galle (1812–1910) during his first attempt on September 23, 1846. It turned out that the observers in England had seen, but not recognized the planet, several weeks earlier. Galileo probably also saw Neptune over 200 years earlier, for it appears on one of his star charts, but he did not recognize that it was a planet.

These discoveries had a profound influence on western thought. It was a dramatic demonstration that the natural laws discovered by Isaac Newton had the power to make precise predictions of the motions of the planets. The universe indeed appeared as well ordered as a clock, behind which perhaps resided the clockmaker.

It has long been known that accretion of planetesimals to construct planets takes too long in the remote outer part of the solar system where Uranus and Neptune now reside. The time needed to form these planets in place exceeds the age of the solar system. Since we see Uranus and Neptune, this model did not meet the modest requirement that the planets already exist. But it has now become clear that they did not form in place but have migrated to their present position.

Current models suggest that their cores formed near the snow line around 6 and 8 AU but were driven out to 10 and 13 AU. They arrived at their current locations of 20 and 30 AU several hundred million years later, due to chance resonant encounters between Jupiter and Saturn as outlined in the Nice Model.

This history of the ice giants explains why is there so little gas in either Uranus or Neptune compared to the abundance in the gas giants, Jupiter and Saturn. The giant planets all managed to form massive cores, by accumulating a myriad of icy and rocky bodies, of which Pluto, Triton, KBOs, TNOs and the Centaurs are surviving examples. Uranus and Neptune thus suffered the usual fate of coming too late to the table, forming a little further out. It is for this reason that these planets are small ice giants compared with the massive gas giants, Jupiter and Saturn. If the gas content of the disk had been larger or had dispersed more slowly, they might have grown to rival Jupiter or Saturn.

Models for the internal structure of Uranus and Neptune suggest the presence of a rock and ice core, covered with an icy oceanic shell that is a mixture of water ice, methane and ammonia. Then there is an outer envelope composed of gas and some ices. However these layers are not clear-cut and they grade into one another. Apart from the fact that Neptune is more massive, there are many differences in heat flow, magnetic fields and density between the pair. The devil is in the detail. Curiously although Neptune receives 40% less sunlight than Uranus, their surface temperatures are similar, for unknown reasons.

Uranus and Neptune rotate at about the same speed. However, they have dramatically different tilts. The axis of rotation of Neptune is tilted, like Saturn, at nearly 30° to the common plane of the solar system. Uranus, in contrast, is lying on its side. Its satellites and nine dark rings rotate about the equator. The only way to knock Uranus over was to hit it with something big. Whatever did the job had to be about the mass of the Earth. It has been suggested that the difference in the heat production inside Neptune and Uranus may be due to different internal structures resulting from this massive collision, just as a boxer might have had his internal organs rearranged from a great blow in the solar plexus.

This pair of icy giants once again demonstrates the difficulties in making similar planets. One might have expected that things might become more uniform on approaching the outer edge of the system, but that hope is not realized. These differences lead us to expect that the commonly found Neptune analogs and super Earths are also likely to differ greatly in detail among themselves. But their abundance among the exoplanets also informs us that nature has little difficulty in forming cores. The problem seems to lie more in capturing gas and so forming giant planets, before the gas is lost from the disks. So we are fortunate to possess gas giants such as Jupiter and Saturn, both of which have played such a crucial role in determining the architecture of our familiar system.

THE OUTER EDGE OF THE PLANETS

Neptune is the true outer boundary of the planetary system, which ends as suddenly as a cliff. This abrupt termination was long a matter for surmising. Initially there were thought to be variations in the orbit of Neptune. These deviations encouraged speculation that there must be yet another major planet lurking out there in the far reaches of the system. Human imagination likes to create hidden monsters.

Thus the idea of the dark planet X arose, apparently so-called by Percival Lowell (he of the Martian canals). Pluto however turned out to be much too small to be a candidate. Alas, the calculated variations in the orbit of Neptune turned out to be spurious. They were due to an incorrect value (by one part in 200) of the estimate of the mass of that planet. When the correct mass, obtained from the gravitational effect on spacecraft passing near the planet, was used, the calculated wobbles in the orbit of Neptune disappeared. The total mass beyond Neptune is less than one tenth of an Earth-mass and probably smaller. We will just have to put up with a solar system that has only eight planets, despite much hope for ten or more, a wish that goes back to Kant. The ancients were content with five, apart from the Earth.

MINIATURE PLANETARY SYSTEMS

There are a host of satellites around the four giant planets. Circling around Jupiter are the four famous Galilean satellites, Io, Europa, Ganymede and Callisto. When Galileo saw in 1610 that these bodies were in orbit around Jupiter, he realized that this provided direct proof of the correctness of the Copernican System. Their orbits around the equator of Jupiter are nearly circular. They resemble a miniature solar system, a point also appreciated by Galileo. These well-behaved satellites are fairly uniform in size and they show a steady decrease in density with distance from the planet, because the amount of rock falls off and that of ice increases in the satellites as one goes away from Jupiter. All these regular features encouraged the view that the Galilean satellites would provide insights into the formation of the solar system, just as one can study the workings of a large machine from a scale model.

Three of the giant planets possess regular satellite systems that indeed mimic miniature planetary systems. Neptune has yet another set of satellites, dominated by Triton. These four scale models provide us with some statistics. Surely we might gain some insights into the formation of planets from studying them.

This hope is soon dashed. The satellite systems of the four giant planets are startlingly diverse. Their satellites are so distinct that they could all belong to different planetary systems without causing surprise.

There are over 160 satellites, but no two are truly alike. Various attempts have been made to classify them, but like many features of the solar system, they mostly defy being placed into neat pigeonholes. However, their names are so enchanting that I make no apology for listing some of them.

The first attempt at a classification puts them into three divisions: Regular satellites, irregular satellites, and a third category of bits and pieces resulting from collisions.

The regular satellites are in coplanar orbits around their parent. The irregular satellites mostly have highly inclined and elliptical orbits far away from the planet. The chief interest in these captured bodies is that they are probably from the early solar system, captured during the migration mayhem. Like fugitives from a lost battle, they were rounded up and captured late in the day.

Some objects refuse to be put into even these three broad pigeonholes. These include Triton, Charon, the Moon, and the tiny satellites of Mars, Phobos and Deimos. Triton revolves backwards around Neptune and is a close relative of Pluto. It is also ironic that our nearest neighbor, the Moon, cannot be placed into any tidy organizational scheme, and indeed needs a separate section to explain how it came to be in our night sky. Our "twin" planet, Venus, adds to the problem by not having a satellite at all. It is clear that the search for some kind of regularities among the satellite systems has failed and that no simple sequence of reproducible events has occurred in the solar system.

Some satellites are worthy of special comment just to illustrate the ability of the solar system to produce unexpected forms.

The Galilean satellites of Jupiter, Io, Europa, Ganymede and Callisto, all display notable differences. Io, tidally heated by the embrace of the monster Jupiter, is in a state of continuous eruption, a Dantean version of hell. Europa, with a salty ocean and icy crust and in resonant interaction with its neighbors, is warmed enough to erase craters on its surface. Ganymede, Europa and Io are locked into resonances that provided enough heat to differentiate Ganymede into a metallic core, silicate mantle and icy crust. Callisto, the furthermost of the Galilean satellites, unlike Ganymede is only slightly differentiated.

The regular satellites around Jupiter accreted at low temperatures, where water ice was stable, without initially undergoing melting and differentiation. Callisto, which is the extreme example, must have formed from rock and ice components without being heated sufficiently to cause early melting of ice. So the Galilean satellites grew slowly in an environment very different from that of the inner planets, which accumulated from smaller bodies through massive collisions that caused melting and separation of cores and mantles.

The Saturnian satellite, Titan [6], another unique body, has always excited interest, both on account of its size and thick atmosphere. It is only a little smaller than Ganymede. Like that satellite, it is bigger than the planet Mercury. The surface temperature is only

95° above absolute zero. The atmosphere is mostly nitrogen with traces of complex organic chemicals. The surface pressure is 50% higher than that of the Earth's atmosphere. It is the only body in the solar system, apart from the Earth, to have stable liquid on its surface. But the liquid is methane, not water, and there is a hydrologic cycle of lakes and rivers of methane, with monsoons of methane under a smoggy orange sky. Titan has an eerie Earth-like landscape. There are deserts at the equator while lakes of methane appear at the poles. Titan is a favorite target for searches for chemical precursors to life or of possible life forms, but what sort of creatures might live in a lake of methane is difficult to imagine.

The strange satellite system around Neptune seems to be mainly a ruin, the result of the capture of Triton, a cousin of Pluto. The arrival of this large intruder on the scene (it is 40% larger than Pluto) must have resembled that of a bull entering a china shop. Any original satellite system was destroyed by collisions. The tiny inner satellites that we now see may be survivors from the debris that remained. Five of them lie close to the rings that surround the planet and are perhaps even broken-up bits left over from the ring-forming event. Faraway Nereid is probably a captured comet. Thus the satellites of Neptune carry a particular message. They show what can happen if a large body is captured. One out of four satellite systems in our own solar system was disrupted. Although the capture of Triton was a random event, it shows how frequently such disasters might occur.

THE COSMIC JUNKYARD

The satellite systems of Jupiter, Saturn and Uranus have so few features in common that they could equally well belong to separate planetary systems. Perhaps the most interesting observation about the small bodies is that there is so little uniformity. The satellite systems are all individual and specific to the planet around which they orbit. There is no grand program for manufacturing them on some sort of cosmic assembly line. The solar system does not look like the product of a well-organized factory for producing planets and satellites. Instead, it looks as though it was assembled from the bits and pieces lying around in some cosmic junkyard.

This gives little cheer to the idea that clones of our solar system exist. If the four giant planets produce such different sets of satellites, we should not be surprised at the differences among exoplanetary systems that we now observe.

PLUTO AND THE KUIPER BELT

Pluto carries much the same sentimental, emotional and historical overload as Father Christmas or the British Royal Family. Even the name for this frozen dwarf, suggested by 11-year-old Venetia Burney, in Oxford, is evocative with its shades of the underworld. To add to the mix, one of man's best friends appeared as Disney's Pluto about the same time. It all makes for a heady cocktail.

Pluto was discovered by accident, a legacy of Percival Lowell's obsession with planet X. In the normal course of events, Pluto might not have been discovered for another 50 or 60 years and so the controversy of whether Pluto is a planet might never have arisen.

The discovery of Pluto's relatives, the KBOs or Kuiper Belt Objects, is reminiscent of the history of the discovery of the asteroid belt. The first, Ceres, located in 1801, was hailed as the missing planet between Mars and Jupiter that was long predicted by the Titius–Bode Rule. But its trivial size and the discovery of Pallas, Juno and Vesta lurking in the vicinity alerted observers that the true situation might be more complex. It was not until 1845 that other bodies were discovered and then, as telescopes improved, a deluge appeared, so relegating the four new planets to the status of minor planets or asteroids. Even the notion that they were remnants of a shattered planet vanished as they shared no common point of origin.

Why then did it take so long to discover the predicted swarm of KBOs beyond Neptune? Probably because Pluto is in such an eccentric and inclined orbit that no one was dedicated enough, or had the funding or sufficient energy to repeat the arduous search technique of Clyde Tombaugh. Although one of Pluto's moons, Charon, was picked up in 1978, it took David Jewitt and Jane Lu until 1992, after 6 years of searching with modern instruments, to discover the next body in the Kuiper Belt, a classic case of the needle in the haystack. Now there is a flood of KBOs that includes three satellites of Pluto.

But Pluto was eclipsed by the discovery in the Kuiper Belt in 2005 of another apparently larger body, Eris, a rival for Pluto as the "King of the Kuiper Belt". This awkward fact ignited the debate over the planetary status of Pluto. Although many thought that Pluto could remain as the ninth planet, by applying the three physical properties of orbital characteristics, mass and roundness, these criteria lead to far too many planets to be practicable.

When Pluto was discovered, the surprise was that it was so small. As data improved over the years, it kept shrinking in size so that it became a joke that by 1984 Pluto would vanish. When its first satellite, Charon, was discovered in 1978, the true radius of Pluto shrank to 1180 km so that it is only 18% of the mass of the Moon and even Triton, its cousin, is much more massive. Something similar is now happening to Eris, which is apparently closer in size to Pluto than originally thought.

Pluto has both an eccentric and inclined orbit and was trapped early into a 2:3 resonance with Neptune, along with a horde of similar bodies, the Plutinos (Figure 14). These are abundant and constitute much of the inner portion of the Kuiper Belt. As Neptune marched outwards they were captured along with Pluto into their resonant orbits which protect them, unlike Triton, from becoming satellites of the planet. Probably none of the KBOs represent a fair sample of the early solar system. Like ancient crones, too much has happened to them since early childhood.

The Kuiper Belt has more than 20 times the mass of the asteroid belt. It resembles a doughnut rather than a belt because many of the bodies are in inclined orbits. Some gaps exist, analogous to the Kirkwood gaps in the asteroid belt, caused by resonances with Neptune. There are two other large populations, firstly the classical bodies, which orbit like the planets in nearly circular orbits with low inclinations (Figure 14). Following the gap of several AU beyond Neptune swept out by that planet, the classical belt extends from about 39 to 48 AU. Its members are probably original constituents of the solar nebula driven outwards by Neptune, that escaped being captured into resonant orbits by the giant. The other group (the scattered bodies) has much more inclined orbits and appears to have a distinct composition. They probably originated near Jupiter and were violently ejected outwards. These two groups are sometimes fancifully labeled cold and hot populations, terms that have nothing to do with temperature. The Kuiper Belt contains perhaps 70 thousand bodies larger than 100 kilometers in size, with perhaps several million bodies of the order of 10 kilometers in diameter, or about the size of Mt. Everest, that mountain of Permian limestone.

The Oort Cloud

Well beyond the Kuiper Belt lies a spherical halo of icy bodies, the source of the long-period comets like Halley. It is termed the Oort Cloud, after Jan Oort (1900–1992), although Ernst Opik (1893–1985) had predicted its existence earlier. Like the Kuiper Belt, it was not an original part of the solar nebula, but was formed as the giants scattered remnants outwards (Figure 14). These icy remnants extend out to over 50,000 AU, almost a light year, although most lie within 2000–5000 AU. The icy bodies are only weakly bound by the gravitational field of the Sun and so are easily dislodged into Sun-crossing orbits, forming the long-period comets. Although the Cloud is estimated to contain a trillion (10^{12}) objects, no members have yet been seen, although Sedna, an icy planetesimal with a highly eccentric orbit from 76 to 960 AU, is a likely inner resident.

Among the many current questions without answers, will other planetary systems have analogs of the Kuiper Belt and the Oort Cloud, or some alternative structures?

A HERD OF CENTAURS

The Centaurs were beasts in mythology that were half man and half horse, just about as strange as their namesakes in the solar system.

They need to be mentioned here as examples of what might be occurring in other systems. The first sighting of these creatures was in 1977, when Chiron was found, wandering apparently alone in the immense void of 10 AU, between Saturn and Uranus. This was a surprise, since it used to be thought that the space between the giant planets was empty, an idea going back to Newton. Chiron, about 175 kilometers in diameter, is one of the most isolated bodies in the solar system. It is a dark gray–black chunk, probably coated, like Comet Halley, with dark tar-like material. Occasionally, it is seen spewing out some gas. This makes it a large comet by definition. Within a million years or so, Chiron will collide with Saturn or perhaps Uranus. It has enough mass to duplicate the beautiful ring system of Saturn, if it were captured into the right orbit. Hopefully, it will not penetrate into the inner solar system on its chaotic path.

The herd of Centaurs may number a few hundred. Probably they are wanderers that were shaken loose from the Kuiper Belt. Like a herd of wild horses, Centaurs have an unstable existence. They will finish up becoming short-period comets, being captured or colliding with one of the giant planets.

COMETS

> "The appearance of comets, followed by these long trains of light, has for a long period terrified mankind, always agitated by extraordinary events of which the causes are unknown. The light of science has dissipated the vain terrors that comets, eclipses, and many other phenomena inspired in the ages of ignorance" [7].

The appearance of comets in the night sky caused much distress in primitive societies, because the starry heavens were thought to be fixed and permanent. Often they were thought to be some kind of disturbance in the atmosphere of the Earth. Even now the appearance of a comet is called an "apparition", a word meaning a ghostly figure or a sudden or unusual, frightening sight. The appearance of the ghost of Hamlet's father in Shakespeare's play meets this definition. Even when not terrifying people, comets were thought to herald great changes. Halley's Comet was visible from April to June 1066 CE, just before the Norman Invasion of England, and the apparition is recorded on the Bayeux Tapestry. The comet was widely taken as an omen, but it was not clear what disaster it portended until the defeat of Harold at the Battle of Hastings in October of that year. Then it was apparent that William The Conqueror's claim to the English throne had heavenly approval. Comets have been held to be responsible for such different phenomena as the paranoia of the Emperor Nero, the collapse of the Aztec empire, religious revivals, and mass suicides (one as recently as 1997 in California).

Many comets are observed each year from the Earth. Once a comet is dragged into a close approach to the Sun, heating drives off the water, other gases and dust, so forming the spectacular tail. Comets remind us of Icarus, who flew too close to the Sun. He came to grief when the wax that held his wings together, melted. Comets last somewhat longer than that unfortunate early flier, but eventually die from the same cause, too close an approach to the Sun.

Comets survive for tens to hundreds of thousands of years. So there has to be a regular supply of these short-lived objects from some distant and plentiful reservoir.

Comets are fragile enough; dirty snowballs as Fred Whipple (1906–2004), who was the first to appreciate their true nature, called them in 1950. Like snowballs, they fragment easily, particularly if they stray within the grasp of a giant planet. Comet Shoemaker–Levy provided a good example. It was captured into an orbit around Jupiter in 1929. For the next 65 years, it slowly spiraled in until July, 1994, when the comet broke up into about 25 pieces due to the gravitational pull of Jupiter, producing spectacular collisions that alerted politicians to the dangers of similar impacts on the Earth.

Perhaps the most striking feature among the comets visited by spacecraft is that they differ widely in detail. Clearly comets are not simple remnants of the nebula. Comets with long periods, such as Halley, originate from the Oort Cloud. Comets are rich in water ice. Although the hydrogen in the water molecule is 99.98% H, a heavier isotope, deuterium (D) is also present in traces. The ratio of lighter hydrogen to its heavier isotope (always expressed as D/H) varies in different environments and so forms a useful tracer for the origin of water. Six long-period comets from the Oort Cloud have D/H ratios twice that of the Earth's oceans, ruling them out as sources of the water on the Earth.

Short-period comets come from the closer-in Kuiper Belt. The D/H ratios of the two measured so far correspond both to that of the Earth's oceans and of the CI carbonaceous chondrites from the outer edge of the asteroid belt. Both Kuiper Belt comets and icy CI meteorites are likely suppliers for our water. This is not surprising as Kuiper Belt comets and icy asteroids from the outer parts of the asteroid belt share a common origin.

The Stardust Mission took samples from the dust cloud around Comet Wild (Vilt) that originated near Jupiter. Although the sample container later crash-landed, scientists were able to salvage some material. Most of the dust was composed of silicates. Many were our familiar solar system material, much of which had been near the Sun. But the most surprising of all was the discovery of some inclusions of refractory minerals (CAIs) formed near the Sun at high temperatures. Clearly, widespread scattering of material had occurred within a few million years of the formation of the solar system. Comets did not turn out after all to be primitive samples of the solar nebula. Like the seekers after El Dorado, we have stumbled across something else.

It has sometimes been suggested that comets come from outside the solar system. As many icy planetesimals finish up being ejected from our planetary system, interstellar space might be heavily populated with free-floating comets. A hyperbolic orbit would reveal comets coming from interstellar space. None have been definitely confirmed so far, but the question remains open.

PLANETARY RINGS

Rings around giant planets are examples of the sort of detail that we can see in our own system and may become discovered among the exoplanets. Accordingly they also need to be mentioned here. The beautiful and exotic rings of Saturn were for a long time the only known examples. They had excited wonder ever since Galileo. He thought that they consisted of two large satellites, one on each side of the planet. In another view, the rings were thought to be lobes, attached like ears to the planet. Then Christian Huygens (1629–1695) discovered that they formed a large ring that was completely separated from the planet. Huygens thought that Saturn's ring was thick and solid. Even as eminent a scientist as Laplace followed this view and thought that it was a solid disk, looking like a sort of celestial phonograph record. Earlier, Kant had correctly supposed the rings to be composed of many small particles, each one circling the planet in accord with the rigid laws of dynamics. Later workers confirmed Kant's notion.

The splendid rings of Saturn have now been resolved in spacecraft photos into many thousands of rings of particles, each circling around Saturn in about a day. Although the ring system is of vast extent (its diameter is 5% larger than the average Earth–Moon distance) the rings are extraordinarily thin, less than 10 meters thick. The average particle size in the rings is only a few meters, although some house and mountain-sized objects are present as well among the rubble. The ring particles are mostly ice, rather than frosty rocks.

The edges of most rings are less than 100 meters wide, kept sharp by small shepherding satellites. The analogy of shepherds (small satellites) controlling flocks of sheep (ring particles) is frequently used. However sheepdogs would be a better term, since they are much more efficient than shepherds, a piece of wisdom inherited from my youth on a New Zealand farm.

Much later other rings were discovered, first around Uranus, then Jupiter and finally Neptune. Compared to the rings of Saturn, the newly discovered rings are mostly darker and much less massive. Jupiter has faint rings, composed mostly of fine dust from nearby moons; Uranus has at least 13 very dark thin rings. Five rings have been identified around Neptune. The amount of material in the

Neptunian ring system is very small. It amounts to about 1% of that of the rings of Uranus, or 100,000 times less than the magnificent rings of Saturn.

What about the inner planets? Searches for rings around Mars have revealed no trace of a ring inside the orbit of Phobos, where one might be expected. Neither are there any signs of rings around Mercury, Venus or the Earth.

The discovery of rings around Uranus, Jupiter and Neptune, as well as the presence of the famous rings of Saturn, naturally raises the question of whether the formation of rings is inevitable during planetary formation. The contrast could hardly be greater between the different ring systems. It is not related to the size or position of the planet. Clearly no uniform ring-making process has been in operation. There is so much variety among the rings, as with everything else in the solar system, that it is difficult to extract any general rules.

What are we to make of all this diversity in mass, structure and composition? There are a few clues. Firstly, there is not much material in the rings. All the material in the rings and shepherding satellites of Saturn could be contained within an icy satellite about 200 kilometers in radius. A small dark body would contain the material in the rings of Uranus. The total mass of the Jupiter ring or of those of Neptune is trivial.

They do, however, share one common property. The rings all lie close in to the parent planet, within about three radii of the planet. This is the famous limit worked out by the French mathematician, Edouard Roche (1820–1883). Any fragile body that comes this close within the embrace of the giant planet will be torn apart. The Shoemaker–Levy comet is a dramatic example. If it had been on a different orbit, it might have formed a dusty ring, rather than the disrupted fragments hitting Jupiter in 1994.

Theory also predicts that the rings should not be stable over long periods. The parent planet should sweep up the particles over periods of the order of a few hundred million years. This is much shorter than the 4.5 billion years that the solar system has been around. If the calculations are correct, the rings have a relatively short life.

The most likely picture is that the breaking up of captured comets, such as Chiron, produces the rings. The tiny particles become the rings, and the bigger bits make tiny moons. These act as shepherds or sheepdogs, according to taste. They also supply dust as they are ground up by collisions among themselves, like boulders in a stream bed. The different masses and colors of the rings are a natural outcome of the break-up of bodies of different compositions, some icy and bright, some dark and rocky.

If the break-up of captured bodies is the correct explanation, then the spectacular rings are latecomers to the solar system. They are accidental features and have no fundamental significance for the origin of the solar system. Thus, like so much else in the solar system, the planetary rings are due to chance events. They form when a small body wanders within the gravitational embrace of a giant planet. The absence of rings around the terrestrial planets is due to their small size. Comets and asteroids mostly collide directly with these planets and explode to form craters. Any broken-up debris is fairly quickly swept up by the planet, rather than forming an extended ring.

Possibly only one ring-making event will occur for each planet over the whole age of the solar system. This model seems consistent with the observations of the totally different rings around each of the major planets. The rings seem to be relatively short-lived, so we must judge ourselves fortunate to be present in the solar system at the same time as the splendid rings of Saturn. Perhaps we should appreciate them even more, since they may well be, like so much else around us, unique in the universe.

THE INNER SOLAR SYSTEM

Our early solar system was dominated by the formation and behavior of the giant planets. But 2 or 3 Earth-masses of rocky rubble were left in the inner nebula after Jupiter had reversed its inward march.

Most of the debris eventually finished up being ejected or in the rocky planets, but some survived to populate the asteroid belt. Thanks to the meteorites delivered to us from there, we now know much of what had happened in the earliest times and so asteroids need to be discussed here.

THE ASTEROIDS

A gap in the orderly sequence of planets between Mars and Jupiter had been noted ever since the time of Kepler. When Johann Daniel Titius came up with his famous rule in 1766, it was clear that a planet was missing between Mars and Jupiter.

Curiously enough, Giuseppe Piazzi (1746–1801), who was checking a star catalogue on New Years Day, 1801, discovered Ceres at 2.77 AU, close to the predicted spot at 2.8 AU, but lost track of it soon after. Ceres, number one in the asteroid catalogue, is the largest of its kind, containing about one third of the mass of the entire asteroid belt. Its diameter is 933 kilometers. Wilhelm Olbers (he of the dark night sky paradox) discovered the next one, Pallas, in 1802. He thought that these tiny bodies were fragments of an exploded planet. Juno (1804) and Vesta (1807) were discovered soon after.

But it was not until 1845 that another asteroid, Astraea, was discovered. The floodgates then opened. Now over 300 thousand have been identified. There are millions of smaller bodies less than a few kilometers across. The orbits of over 100 thousand asteroids are well enough known for them to be assigned numbers. The asteroids have been a great source of irritation to astronomers, who are mostly interested in more distant objects. This swarm of bits and pieces gets in the way, leading to the insulting title of "the vermin of the sky".

Although in popular mythology, the asteroid belt is portrayed as a swarm of boulders that would be a hazard to intrepid space travelers, they are mostly separated by millions of kilometers and don't add up to much. The total asteroid mass amounts to only 5% of that of the Moon. Asteroids are mostly spinning with rotation rates of a few hours, due to collisions. Many have satellites. Many are found in the Main Belt, lying between 2.1 and 3.3 AU although there are several gaps called Kirkwood Gaps, after Daniel Kirkwood (1814–1895), an Indiana astronomer. The gaps occur at simple ratios of the orbits of the asteroids with Jupiter. At these resonant locations, the gravitational force of this giant ejects some of them in our direction, so providing us with meteorites.

This is the most valuable contribution that the asteroids make to our understanding, since the meteorites provide us with samples of the early solar system. We can discover both their composition and age in terrestrial laboratories. However, there are a large number of asteroid compositions not recognized among the present collection of meteorites. We keep finding new varieties in Antarctica and in deserts. So we should be cautious in placing too much reliance on the present meteorite collection as complete.

Worthy of mention among the asteroids are the Trojans, that lie in the same orbit as Jupiter. Like camp followers keeping well clear of danger, they occupy two stable positions, L4 and L5, 60° ahead of and behind the orbit of the planet. These positions are named for the great French mathematician, Comte Joseph Louis Lagrange (1736–1813). There are also Trojans, possibly numerous, around Neptune, captured during the outward march of the giant. A few exist around Mars and even the Earth has one. Saturn and Uranus, lacking stable gravitational positions, seem unlikely to possess Trojans.

Sunwards of the Main Belt, at about 2 AU and separated from it by the hole caused by the 4:1 resonance with Jupiter, is a group called the Hungarias. They mostly have high inclinations and are probably the source of the E-type meteorites, in which all the iron has been reduced to metal.

Of more immediate concern to us are the Near-Earth Asteroids (NEAs) whose orbits cross those of the inner planets. They are divided into the Apollos, Atens and Amors. The Apollos, despite their lovely names, have the capacity to do us deadly damage, as they are in Earth-crossing orbits. The Atens have orbits that lie within that of the Earth, but cross our orbit at their maximum distance from the Sun. The Amors have orbits that cross that of Mars and approach, but do not cross, that of the Earth. The largest one, Ganymed, not to be confused with Ganymede, the big satellite of Jupiter, is over 38 kilometers in diameter.

Although there are many varieties, asteroids form three main types. S-type asteroids, the source of our most common meteorites, are found in the inner parts of the belt. Some have been melted and differentiated, like Vesta. Related to these are the M type, that are likely the iron–nickel cores of such bodies, remnants from collisions. These are quite common, perhaps forming 10%.

The other, and for our purposes the most important, are the C and D types, with a few percent carbon and water. They are distinct from the S type, show some similarities with comets and have likely come from beyond Jupiter. These are the main components of the outer parts of the asteroid belt beyond about 2.8 AU, constituting around 75% of all asteroids. They are the most important for our story, as the composition of the most primitive, the CI carbonaceous chondrites, matches that of the Sun for the non-gaseous elements (Figure 3).

Apart from visits by spacecraft, much of the information about the surfaces of the smaller bodies comes from studies of the reflectance spectra of asteroids in the visible and near infrared parts of the spectrum. As only the upper few microns of the surface are probed by this technique, these data record the effects of the solar wind, sputtering and micrometeorite bombardment of airless surfaces.

Attempts to use reflectance spectroscopy at first appeared to be a relatively straightforward task. The most abundant meteorites to fall on the surface of the Earth are stony meteorites, the "ordinary chondrites". Such meteorites that were photographed during their fall through the atmosphere and subsequently found on the ground pointed to their origin from within the main asteroid belt, so it was a matter of comparing the laboratory spectra of ground-up meteorites with those obtained telescopically from asteroidal surfaces. This approach was reinforced by the success in identifying 4 Vesta as the source of the eucrite class of basaltic meteorites.

But nature is subtle and a paradox soon arose. The identification of their parent bodies by comparing laboratory with asteroidal spectra, apparently successful in the case of the eucrites, proved elusive for the abundant chondrites. In spite of extensive searches, no suitable parental asteroids appeared among the thousands of candidates. Although the abundant S-type asteroids, which dominate the inner part of the main asteroid belt, appeared to contain the appropriate mineralogy (metal, olivine, pyroxene), the absorption bands in the asteroidal spectra, compared with the meteoritic spectra obtained in the laboratory, were much weaker and redder.

This problem was resolved by the return of minute grains from the tiny asteroid Itokawa. The samples revealed that the impact of solar wind protons (mainly hydrogen) on minerals had produced nanoparticles of metallic iron that coated a few microns of the surface. This so-called "space weathering" had altered the spectral appearance. But the data also revealed that the asteroid was eroding "rapidly", losing several centimeters of its surface every million years, and so Itokawa is likely to vanish in a few hundred million years.

It also became clear from telescopic observations that 4 Vesta had suffered a relatively recent large impact that had produced both the eucrites from Vesta and a relatively fresh and un-weathered surface on the asteroid. So the dilemma was resolved and the ordinary meteorites indeed come from the plentiful S-type asteroids.

How did the asteroid belt acquire its present form? The swarm of tiny asteroids is there because of the early formation and inward march of giant Jupiter. About half were thrown right out of the solar system. Perhaps a quarter finished up in the Sun, while a similar fraction hit the Earth or Venus. Like a victorious army scattering a defeated foe, Jupiter then altered the orbits of the impoverished remainder. They were scattered so that they could not reform ranks and collect themselves into a planet. Like any fleeing crowd, many collisions took place between these survivors.

The differentiated S type dominate the inner belt, while the outer belt has the very distinctive more primitive C type. Melting of water ice, fortunately without changing their chemical composition, has altered some of the minerals, including the famous Cl class.

Much intellectual effort has been expended in trying to account for this zonal order. Heating from the early Sun was shown to be inadequate. Zonal ordering of short-lived radioactive species that might heat the inner but not the outer belt proved equally unlikely. Many other models arose.

A more likely current model explains many of the attributes of the asteroids. The S-type asteroids originate in the inner nebula, within 1 to 3 AU of the Sun, and are melted by heating from short-lived radioactive isotopes, and collisions. During the migration of the giant planets, the C and D types, originally icy planetesimals from the outer parts of the disk, were scattered inwards as well as outwards. They populated the outer parts of the asteroid belt. Some traveled in as far as the rocky planets, providing volatile elements and, importantly, water. So the C- and D-class asteroids come from afar. The largest asteroid, 1 Ceres, may be an example. The outer zone carbonaceous chondrites thus have a similar source to the Jupiter belt comets, so that there is no clear distinction between them.

Vesta, a differentiated asteroid [8]

4 Vesta is the third largest of the asteroids and has been visited by the Dawn Mission. The most interesting feature of Vesta is that it forms a scale model of a terrestrial planet with a metallic core, a mantle and a basaltic crust within a few million years of T_{zero}.

In some regions, deep cratering has exposed the mantle. The surface is heavily cratered with the south pole area occupied by an enormous central peak crater (Rheasilvia), 500 km in diameter and 19 km deep, formed by the impact of an 80-km-diameter asteroid. The impact is estimated to have occurred about 1 billion years ago. The much-battered "central peak" is 15 km high and 200 km in diameter. This crater is close to the diameter of the asteroid and exposes mantle rocks as well as providing a coating of fresh rocks over much of Vesta. The impact excavated mountain-sized chunks into space. As if riding on a conveyor belt, they are slowly drifting towards the 3:1 resonance with Jupiter at 2.50 AU. Pieces from that location are the source of the meteorites that originated from Vesta. These include the eucrites which are basalts, the diogenites that are derived from deeper in the mantle and the howardites that are mixtures of both.

Vesta accumulated from dry planetesimals that were already strongly depleted in volatile elements. The asteroid melted during accretion or shortly afterwards, from heating due to short-lived radioactive isotopes of iron or aluminum, assisted by impacts. This interesting asteroid has preserved a differentiated crust over the history of the solar system. However, it has been subject to too many collisions to preserve much evidence of volcanic activity on its surface.

There have been many claims for bodies thought to be analogs for conditions on the early Earth and Mars. The basaltic crust of Vesta appears to be the best analog for the early crusts of Mars and the Earth. Other options such as the volcanically active crust of Io or the present basaltic crust of Venus, although often suggested, seem less likely.

The chemical composition of the basaltic eucrites bears a very close resemblance to that of mare basalts from the Moon. This similarity extends to a near-identical depletion in siderophile and volatile elements. For example, the K/U ratios of the eucrites overlap the low values recorded in the Moon (Figure 4). So the asteroid Vesta underwent a similar sequence of loss of volatile and siderophile elements to that recorded on the Moon. The angrite class of meteorites which come from a probably similar asteroid, have even lower volatile element abundances. Comparable processes produced many dry volatile-depleted "igneous" planetesimals in the inner nebula. These eventually accreted to form the terrestrial planets. Probably the body (now named Theia) that hit the Earth and formed the Moon was one of these.

Meteorites

In 1492, famous as the year that Columbus discovered America, a meteorite fell near the village of Ensisheim in Alsace. This region is now a province of France, although it then formed part of the Holy Roman Empire. The fall was spectacular. The explosions as the body broke up in the atmosphere were heard for hundreds of kilometers, as the body travelled northwestwards across Switzerland and exploded over the Rhine. The Emperor took this event as a favorable omen from Heaven and made a successful attack on the French. He ordered that the stone be preserved and it is still there in the Town Hall. The villagers put a chain around it, to prevent it flying back up into the sky. They also attached to it the following notice that is still relevant: "Many know much about this stone, everyone knows something, but no one knows quite enough" [9].

However, we do know much about the history of the early solar system, only because of meteorites [10], which is why they are discussed here. A decisive advantage is that we can analyze and, importantly, determine the ages of meteorites, a non-trivial task that can only be accomplished in terrestrial laboratories. Some of the rare components of meteorites, such as diamond and silicon carbide, retain a memory of the time before the solar system existed. The abundances of elements in the CI class provide us with the composition of the dust or rock component in the primordial nebula. Some meteorites come from small asteroids that melted and produced volcanic lavas within a few million years of the beginning of the solar system.

The number of attempts to classify them illustrates their diversity: new systems appear every few years. There is broad agreement, however, that most fall into the three traditional categories: irons, stones and stony irons.

The spectacular masses of iron, which we see in museums, are everyone's idea of a meteorite. These come from the metallic cores of at least 60 planetesimals. These were around 100 km in diameter and formed, melted and were broken up within 1 or 2 million years after the beginning of the solar system. The formation and melting of planetesimals was probably caused by the presence of short-lived radioactive isotopes of aluminum or iron, assisted by heating during collisions. The formation of the iron meteorites was one of the earliest events in the solar system. Dynamical studies suggest that they formed near the Sun.

The most common stony meteorites are called chondrites, of which about 25 classes have been distinguished. They contain abundant tiny spheres, typically about a millimeter in size. These are the famous "chondrules" discovered over a century ago by Henry C. Sorby (1826–1908). He was a gentleman scientist in the Victorian tradition, who invented the powerful technique of examining thin transparent slices of rock under a microscope. When he turned his attention to meteorites, he recognized that the chondrules had been melted droplets that had cooled mostly to glass. We have not made too much progress since, although human ingenuity has suggested all possible processes without much agreement. Theories trying to account for the origin of chondrules fill many multi-author volumes.

The formation of chondrules seems to have spread over a few million years, overlapping with melting and differentiation of asteroids into metal cores and rocky mantles. The chondrule factory must have been efficient, for at least half of the material in the common meteorites (or "ordinary chondrites" in the jargon) seems to have passed through it. It has become clear that chondrules had been dust balls in the nebula that were flash melted. The region in which they formed could not have been hot, as they had to melt and cool down within minutes to hours. Curiously the melting was selective. The chondrules are mostly silicate that had already been depleted at T_{zero} in volatile elements, while the metal and sulfide grains mostly survived the process and now form the fine-grained matrix of the meteorites. The current best bet for the melting process is that shock waves, presumably generated from the early Sun, continually swept through the cool inner nebula, but there is no agreed process for producing them.

What we do know is that they formed about 2 million years after T_{zero}. This date, 4567 million years ago, marks the earliest measurable event in our solar system. It records the formation of high temperature minerals that are found as small refractory inclusions in most meteorites. These sometimes reach centimeter dimensions and are usually called CAIs (for Calcium–Aluminum Inclusions). They have been subject to repeated cycles of extreme (2000 K) temperatures very near the early Sun and have then been sprayed out over the nebula. As noted above, some have even been found in comet dust.

The CI carbonaceous chondrites, from far out in the asteroid belt, provide us with the primitive composition of the rocky or dusty component of the solar nebula, as already discussed. All the other groups of stony meteorites from closer in have lost the ices and significant amounts of the volatile elements. They are dry, containing none of the water-bearing minerals. As this depletion is also found in the rocky planets, this important information tells us that the inner nebula, out to about 3 AU, was depleted relative to CI and the Sun. This event is known to occur from isotope studies close to or perhaps at T_{zero}.

Some of our most striking meteorites, the stony irons, are mixtures of metallic iron and the green mineral, olivine. These came from the boundary between the metallic core and rocky mantle, as the asteroids were broken up in collisions that sent the bits and pieces tumbling Earthwards. Others are basalts, like the eucrites from Vesta.

We also have meteorites from the Moon and Mars. As we have samples from our satellite, there is no doubt about the source of the lunar meteorites. The definitive evidence that meteorites also come from the red planet is that they contain trapped gases that match the composition of the thin Martian atmosphere which was measured by the Viking Landers in 1976.

MERCURY [11]

Mercury is an excellent example of the unexpected diversity to be found among rocky planets. The data from the Messenger mission showed that all previous ideas about the origin of Mercury were wrong, just as was the case with the Moon. Once again, the power of small amounts of data over theories was demonstrated.

There are a number of unique features about Mercury. It might have been supposed that a tiny planet closest to the massive Sun would have the most regular orbit in the solar system. But the orbit is tilted at over 7° to the Earth–Sun plane and is the most eccentric ($e = 0.21$) of all our planets, possibly a consequence of collisions. Being close to the giant Sun, the spin of Mercury has become locked in by its giant neighbor. Mercury's case is a bit different from that of the Moon. It rotates three times for every two orbits around the Sun, ensuring that its entire surface gets to be baked at temperatures hot enough to melt lead. The floors of some deep craters at the poles are shielded from the glare. As there is no effective atmosphere, these deep holes remain below the freezing point of water. Earth-based radar reflections have long indicated some trapped ice, now confirmed by the Messenger mission.

Nevertheless, the surface has about 4% sulfur, illustrating, as does the innermost Galilean satellite, Io, that it is difficult to evaporate volatile elements even from small bodies. The high density of the planet implies a large metallic core that constitutes 85% of planetary volume. An unexpected discovery was that Mercury has a magnetic field. It is very much weaker than the field for the Earth, but presents us with some interesting dilemmas. The magnetic field of the Earth is thought to form as a dynamo in a liquid metal core. But Mercury is so small that an iron core would have frozen aeons ago. Probably the core of Mercury contains enough sulfur to keep the upper part molten, so it is clearly a different beast from Venus that has no field. Magnetism is almost as much of a puzzle now as it was when William Gilbert (1544–1603) wrote his classic text *Concerning Magnetism, Magnetic Bodies and the Great Magnet, Earth,* in 1600.

Another unique feature of the planet is the presence of large fault scarps that cut across the surface. They are about one kilometer in height and several hundred kilometers long. They tell us that the planet has shrunk by somewhere between 1 and 2 kilometers in radius. Much of the contraction is due to the cooling and solidification of the mantle and crust around the large core. The scarps cut the older craters and plains. Younger, fresher craters cut the scarps and some of the smooth plains are younger. This shrinking thus occurred towards the end of the massive bombardment and must have happened, from the cratering record, before 4000 million years ago. Following this initial contraction of the planet, the radius of Mercury has been unchanged for around 4 billion years.

The oldest visible surface is heavily cratered, covered in part by plains and followed by a later cratering event, of which the Caloris Basin, 1500 km in diameter, is the most dramatic example. There is much evidence of volcanic activity that includes individual volcanoes and extensive smooth plains, where fluid lavas have erupted. Although these plains were often thought to be ejecta sheets formed during impacts, like the Cayley Plains on the Moon, they are the result of floods of fluid lava. But these are too Mg rich and Fe poor to be basalt.

There are only minor differences in color across the Mercurian surface, so that there are no dark lava plains like we see on the Moon. The surface composition is low in iron and titanium, very different from the Moon, and the lavas are high in magnesium. The nearest analogs on Earth are Mg-rich lavas, called komatiites, which erupted here mainly during the Archean.

Perhaps the most unexpected result from the Messenger orbital data was that the surface composition has about 1150 ppm potassium, 220 ppb thorium and 90 ppb uranium (with errors of 10 to 20%). These element concentrations on Mercury resemble those on the Earth, Venus and Mars, telling us that Mercury has a similar volatile element composition to the other rocky planets, much higher than those on the Moon or the asteroid Vesta.

Theories for the origin of Mercury need to account for the large size of the metallic core and the only 400-km-thick silicate mantle wrapped around it. The high density of Mercury has had a fatal attraction for modelers of grand unified theories for the origin of the solar system. Early attempts to explain the composition of Mercury, and in particular its high iron abundance, were seduced by its proximity to the Sun. This was because the planet apparently anchors one end of a sequence extending from the high-density inner planets out to the low-density outer solar system bodies. This variation appeared to be consistent with a decrease in temperature outwards from the Sun.

Most models fell into two classes. One suggested that high temperatures had evaporated over half of any original rocky mantle. This would result in the removal of any volatile elements. However, the high surface abundance of the volatile element, potassium, rules out such notions of removing the silicate mantle by such evaporation.

The alternative model to account for the strange composition of tiny Mercury is that much of the rocky mantle was lost during a massive collision. Current estimates are that the impacting body was about one fifth of the mass of Mercury, hitting the planet head-on at 20 kilometers per second. In this scenario, the initial mass of Mercury would be about twice its present value. The rocky material would be smashed into pieces less than a centimeter in size, and mostly swept away to finish up in the Sun, Venus or the Earth. The tough metal core hung together and wrapped itself, like a cosmic beggar, in a thin cloak of rock.

A routine objection to this model is that the high temperatures in such a collision would evaporate the volatile elements, something that has clearly not happened. Contrary to popular wisdom, collisions, although they may alter metal to rock ratios, do not cause evaporation of volatile elements. These elements were depleted at T_{zero} by solar winds in the inner nebula, as described in Chapter 2, not by later collisions.

This is well illustrated by the composition of the asteroid 4 Vesta, that formed within a few million years of T_{zero}, but which resembles the much later-formed Moon in its depletion of volatile elements. Like Vesta, Mercury formed early, but with a more typical

Mars-like budget of volatile elements; a consequence of just what happened to hit the growing planet.

This model also provides for the strange orbit of Mercury. A bigger collision might have left us with only the iron core and so an iron planet at which to wonder. Mercury displays both similarities and extreme differences from the Moon, both bodies being products of massive collisions. This again points up the random nature of events during the early history of the solar system.

Thus, Mercury is another battered survivor from the turbulent early history of the solar system. Many similar-sized bodies existed in our region before the final sweep-up, with most of them finishing up in Venus or the Earth. However, Mercury with its high density is so strange that it falls into a special category, unique even by the standards of the solar system. This tiny planet is another example of the refusal of the planets and satellites to be put into neat pigeonholes.

MARS [12]

Among all the planets, Mars has a unique fascination for the human imagination. This is because the surface conditions on this planet are closer to those on the Earth than on any other body in the solar system, although it has two tiny satellites, Phobos and Deimos, that are probably captured asteroids, and the planet lacks a large stabilizing moon. Mars is a cold desert. The average temperature is 55° below zero centigrade, or 218 K, although it rises to over 20°C in the summer near the equator. Although it is only 11% of the mass of the Earth, we could imagine living there. Even the day on Mars is close to that of the Earth, while the tilt is currently close to that of the Earth. It would be more uncomfortable than living at the South Pole, with the snow blizzards replaced by long-lasting dust storms. A traveler would need oxygen, water, shelter and a good source of energy.

The Romans identified Mars with the God of War, from its red color, and it has exercised a strong hold on the human imagination. Mars has been a favorite location for science fiction and the imaginary inhabitants of the red planet usually have had undesirable characteristics, even by our standards. The invasion of the Earth by aggressive Martians in *The War of the Worlds* by H. G. Wells (1866–1946) was one of the first examples. Late in the nineteenth century, Giovanni Schiaparelli (1835–1910) claimed that regular channels crossed the surface of Mars. These he called canali, Italian for channels. Percival Lowell (1855–1916) took them to be canals, with vegetation along their banks. He finally mapped 437 of them crossing the Martian surface. His notion was that they had been constructed by a civilization to bring water from the ice caps at the poles to the parched equatorial regions. The white polar caps do in fact contain permanent water ice, which gets covered by a seasonal frost of carbon dioxide, our familiar "dry ice".

The apparent presence of the canals excited much interest, as they would have constituted evidence for the existence of a technically advanced civilization. The problem with observing them was that they were at the limit of the resolving power of telescopes in use over 100 years ago. Thus they were seen by some observers but not by others. There was no question that the canals were produced by intelligent beings. The significant question is on which side of the telescope the intelligence was located, as the famous canals turn out to have been optical illusions. The Martian canal story is a cautionary tale for scientists, illustrating the problems of interpreting data close to the limits of resolution.

Free water cannot exist at present on Mars. In spite of endless rediscovery, it has been known for decades that there is water on Mars, most of which is now locked beneath the surface as permafrost or in the polar ice caps. In contrast to the boulder-strewn rubble that surrounds craters on the Moon, those on Mars look slushy, like the splash from a pebble dropped into mud. The heat of the impact melted the subsurface ice and a muddy mix of rock and water was thrown out.

A visitor would have to protect themselves from the dust storms. Unlike those in deserts on Earth, they can last for months. This is due to a combination of the very fine size of the dust grains, typically a few microns, and the low Martian gravity. The large dust storms are responsible for the changes in the surface that are visible in Earth telescopes. These dark and light patterns of the surface were first thought to be due to seasonal changes in vegetation. The dust is typically about a meter thick on the surface and arises mainly from the great impact basins in the south. As the dust settles back onto the surface, it forms spectacular dark bands within the seasonal layers of ice at the poles.

Formation and geology

The growth of Mars was stunted as the neighborhood had been impoverished by the nearby giant Jupiter. The planet has a lower uncompressed density than the Earth, but the silicate portion is enriched in iron and moderately volatile elements compared with the Earth. Isotopic evidence suggests that Mars formed within a few million years of T_{zero}. Like Mercury, Mars is thus a survivor from the earliest days of the solar system. Many other Mercury- and Mars-sized bodies formed within the inner solar system, but were later swept up by Venus or the Earth.

Like the Moon, Mars melted early, probably forming a magma ocean, a sulfur-rich core, a silicate mantle rich in iron and a crust of lava. There is no evidence for the presence of a crust on Mars like the continental crust of the Earth.

Geologists have divided Martian time into four great periods, Pre-Noachian, Noachian, Hesperian and Amazonian, from oldest to youngest. A table discussing these may be found in the Appendices. The Noachian of much interest is very ancient, dated approximately between 4100 and 3700 million years ago. The Pre-Noachian before that period, corresponds to our Hadean. Dates for the two younger epochs are less certain but their boundary lies at around 3000 million years, deep in our Precambrian. Thus the Amazonian period on Mars corresponds to most of our geological record.

Both static landers and rovers have investigated the surface of Mars which is dominated by basalt. The crust is perhaps 30 to 50 km thick with over 30 km of relief from the top of Olympus Mons to the bottom of the great chasm of Valles Marineris. The Martian crust is mostly ancient but volcanism has persisted for most of Martian history.

The crust in the northern hemisphere, lying under a pink sky colored by the red dust, consists mainly of monotonous lava plains. These cover a primitive cratered surface. The crust in the south lies between 1 and 3 kilometers higher than the rolling basaltic plains to the north. This southern crust is also heavily cratered but the craters are visible. Although oceans covering the low-lying northern plains have frequently been proposed, evidence such as former shorelines and the like has evaporated as better topography and photography have become available. The origin of the difference in elevation between the north and south remains unresolved, but is probably related to convection deep in the mantle early in the planet's history.

A large bulge on Mars, the Tharsis plateau, has been the site of volcanoes for nearly 4 billion years. This protuberance puts Mars "off balance". The combination of this and the absence of a large moon allow shifts in the tilt of Mars by up to 60°. Tharsis is the surface expression of a hot plume rising in the mantle. Mars is a one-plate planet and so the hot spot remained in one location, building a giant edifice. Suggestions of a very early phase of plate tectonics persist, but the evidence is unconvincing and Mars has been a one-plate planet with a very different geological history from that of the Earth, Venus or Mercury.

Mars makes up for being tiny by producing the largest landforms in the solar system. Among other evidence, this tells us that geological evolution is unique to each planet, although the operation of the processes of geology may produce similar effects. The largest mountains on the Earth appear as mere pimples compared with mighty Olympus Mons. This monster volcano rises to an elevation of 26 kilometers above the surrounding plains and spreads over a distance of 600 kilometers. The great valley of Mars, Valles Marineris, is 4000 kilometers long. It is so wide that the far rim would be beyond the horizon to a visitor standing on the near side. The Grand Canyon of the Colorado could be dropped out of sight into it. But unlike that river-cut canyon, Valles Marineris is a rift valley, caused by the crustal stresses imposed by Tharsis. Although nearly 4000 million years old, it has still not been filled with sediments, unlike such depressions on Earth.

Like the Earth, Mars has sedimentary rocks, but there are major differences between the two planets. On Earth, many sedimentary rocks are formed from the extensive weathering and erosion of the granitic continental crust, driven mainly by plate tectonics, voluminous water and carbonic acid. Clays, quartz and carbonates are common products.

Mars, in contrast, has a crust of basalt. Water has been scarce. Large amounts of sulfur and chlorine are exhaled by the long volcanic activity. There is no oceanic sink for these volcanic gases, so sulfuric and hydrochloric acids form by combining with the traces of water. Acidic mists attack the basaltic rocks, a different scenario to that on Earth. But it's a slow process so that rocks billions of years old often show only a thin altered crust.

The message is that weathering and erosion, so common on the Earth that old rocks are hard to find or interpret, is painfully slow on Mars. The sediments found by the Mars Exploration Rovers (MER) were about 3500 million years old (Figure 15). Sulfates formed from the alteration of the basalt are everywhere, even in these very old rocks, but clays are less common except in the Noachian.

On Earth, clays are formed at or near the surface. However on Mars, in contrast, the Noachian clays formed below the surface by reactions with subsurface water. They do not constitute evidence for early wet surfaces. Free silica, liberated by weathering from the surface basalt flows, is common. This confuses superficial examination leading to notions that rocks more evolved than basalt might be common, as in the crust of the Earth. As might be expected in an acidic environment, sulfates are common but carbonates are rare. They are found in only a few locations on Mars, possibly formed by local

FIGURE 15 Burns formation, Mars

Planets record their history. Ancient sedimentary rocks, about 3.7 billion years old, at Meridiani Planum on Mars, photographed from the Opportunity Rover. The Burns formation is composed of cemented sandstones. They were originally sand dunes in a desert. When groundwater reached the surface briefly, temporary small inter-dune playa lakes formed. See text. (Courtesy Scott McLennan.) See also color plates section.

alteration by less acidic water. All this contrasts strongly with the benign situation on the Earth, with its vast deposits of limestone, as in the famous "White Cliffs of Dover".

Large volumes of sediments have been transported on Mars mainly by the action of wind. Kilometer-thick sequences are found. This should not be surprising. On Earth, the last ice age, spanning about 3 million years, produced kilometer thicknesses of wind-driven loess in suitable locations. On Mars these processes have been going on for over 4 billion years.

The sediments encountered by the Mars Exploration Rovers, notably Opportunity, that found the Burns Formation at Meridiani Planum, were over 3500 million years old (Figure 15). At this location, there once had been a desert. First to form were sand dunes in the sequence of strata exposed. These were next covered by sheets of sand through which groundwater rose, saturating the rocks and forming globules of hematite, the famous "blueberries". Water finally oozed out on the surface forming shallow lakes between the dunes. Such "playa" lakes are common in deserts on Earth. Finally, the drifting sand overwhelmed them. This is a typical desert scenario on Earth, except that this happened over 3500 million years ago on Mars. In summary, the exploration of Mars tells us that surface water has appeared only briefly and the planet has always been a cold, arid desert.

Why is there so little water on the planet? The D/H ratio of water in the Martian atmosphere is higher than that of the Earth, probably due to evaporation which removes H, the lighter isotope. But water is not a given in rocky planets. It was originally supplied by the random accumulation of asteroids from the outer belt and comets from beyond Jupiter. In this accidental process, Mars just happened to obtain much less than the Earth.

An early warm wet Mars?

> "the final blow to the possibility that the surface of Mars was a 'warm and wet' incubator for any martian life" [13].

The concept of an early warm wet Mars, complete with tinkling streams and primitive life, is one of the staples of science fiction. How did this notion arise and is there any scientific evidence for it? Was there an early wet warm period on Mars in which life itself might have begun?

The widely cited evidence from clays has been disproved as noted above. But on the oldest parts of the crust, among the many Noachian craters that tell us the surface is about 4000 million years old, there are many valleys that bear a superficial resemblance to river valleys on Earth. It is generally agreed that they were eroded by water, unlike the rilles on the Moon that were lava channels. But the resemblance to the stream patterns on Earth is more apparent than real. The Martian valleys most closely resemble those produced by flash floods in very dry deserts on Earth.

Valleys on Earth are cut by rainfall. So did it rain on Mars in that remote epoch? One common solution to producing rain to erode the valleys was to call on an early "greenhouse" that formed a thicker, warmer atmosphere. Many models have been proposed to form such a greenhouse atmosphere on early Mars, but none work. The problem is even worse, because of the famous "faint early Sun" problem. The Sun was about 30% less bright at that distant time, so Mars would have been even colder than now.

Many other models have been suggested. Water trapped in the early crust as permafrost might be melted by volcanic heating. As it seeped out, it could cut valleys by headwater sapping, a process familiar on Earth, where valleys form by water oozing out in springs that undercut the rocks. This process produces amphitheatre-headed valleys, like those on Mars. But these features are also characteristic of valley heads formed in flash floods.

A current model relates the formation of the valleys to brief periods of intense rainfall following meteorite impacts that are common in the Noachian. Following a collision, the entire planet was shrouded by hot debris from the massive strike. The craters formed by the impact were surrounded by an atmosphere of steam. Apart from water present in the asteroid and impact site, much came from melting of permafrost and the ice caps. Such wet climates lasted for decades following a massive impact. As the number of impacts declined in the Late Noachian, so the formation of these river-like valleys ceased [14].

Estimates produce up to 600 meters of rainfall from the steam atmospheres formed during these impacts of meteorites, comets or asteroids. This water not only produces the valley networks, but also erodes the surface. Estimates of the erosion on Mars, based on the degrading of craters and the amount of sediment on the surface, have called for between 50 and 500 meters of rain. But all of these calculations have orders of magnitude uncertainties, a common situation on Mars.

The outflow channels

Massive channels that run for hundreds of kilometers are the most dramatic evidence of the past presence of water on the Martian surface. They have nothing to do with Lowell's canals or the valley networks discussed above. They are typically about one kilometer deep with streamlined walls and are up to 100 kilometers wide. In the midst of the channels there are kilometer-sized islands shaped like large teardrops, eroded by the rushing torrent. The consensus is that water cut the channels, although other liquids such as fluid lavas, glaciers or even carbon dioxide, have been suggested. Only massive floods on a scale that Noah might have recognized can cut such channels. The channels appear to start instantly in areas showing chaotic collapse. This looks like a signature of internal causes. Probably volcanic heat suddenly melted a lot of subsurface ice. The water rushed out as a catastrophic flood that reached many times the volume of the Amazon River on Earth. Then it sank into the ground or evaporated away just as quickly.

My interpretation of the Martian landscape is that Mars never had an early Earth-like climate, but was always a frozen desert, much drier than the Atacama Desert in Chile. But it was subject to temporary periods of heavy rainfall, generated by meteorite, asteroid or cometary impacts, that cut the valley networks. Later flash floods, initiated by internal volcanic heating, swept across the surface carving out the colossal "outflow" channels. Some water remained trapped as permafrost or in the ice caps, while the rest evaporated.

Life on Mars?

Mars is fascinating to us since it is the only other site in the solar system that has some approach to Earth-like conditions. The subject, littered with canals, princesses and alien invaders of the Earth, has long captivated human imagination and has been a favorite of science fiction.

A major reason for sending the NASA Viking Landers to Mars in 1976 was to attempt to discover whether life was present on that planet. Unfortunately, unanticipated conditions on the surface defeated this attempt to make exobiology a science that contained some subject matter. Although the Viking Landers carried three experiments based on biology, all gave ambiguous results. Their strange results have generally been attributed to the presence of a strongly oxidizing component in the soil. These experiments illustrated the extreme difficulty in designing tests to identify life outside the Earth.

A secondary lesson from the Viking Landers is that one should not make the experiments on board unmanned spacecraft too complicated. The experience from the biological experiments on the Viking Landers was that if you did not get the expected answer, you received data that could not be understood. Examination of a few grams of the Martian soil, returned to Earth, would have quickly told us what was causing the strange reactions in the biological experiments.

In fact, a decisive experiment on the Viking Landers was carried out by a mass spectrometer designed to identify organic compounds. Nothing was found down to parts per billion levels. In the total absence of carbon, no terrestrial life forms can be expected. This contrasts with the surface of the Moon, on which even a few parts per million of organic compounds are present. This may not sound very much, but it is over 1000 times greater than in the soil of Mars. As such, molecules must also be added to Mars by meteoritic or cometary infall; this is an apparent paradox. This puzzle has been partly resolved by the discovery of strongly oxidizing compounds that would destroy organic molecules in Martian soils. The question of life on Mars, following the negative results from the Viking Landers in 1976, was summarized by a cartoon of the period: "It's been a disappointing year! Nothing on TV, nothing in Loch Ness and nothing on Mars".

This search was given an extra impetus in 1996 by claims of traces of early bacterial life in a meteorite from Mars. This rock, now famous as Allan Hills 84001, was excavated from beneath the surface of Mars by a meteorite impact 16 million years ago. It eventually fell to Earth in Antarctica 13,000 years ago, from where it was collected from its icy bed in 1984. The rock itself is very ancient, 4.1 billion years old, far older than any of the other meteorites from Mars. Several lines of evidence were claimed in support of the notion of primitive life which, taken together, might indicate the presence of some ancient biological activity. Chains of evidence are notorious for weak links. None of these claims have been substantiated by other workers, so the "smoking gun" remains missing.

The search for life outside the Earth has now been refined to "follow the water", although the presence of water on Mars has been endlessly rediscovered ever since it was identified decades ago in the polar ice caps. But not all water on Mars is suitable or hospitable to life. There is much evidence from Mars of water that is not only acidic, but too salty for even the most salt-tolerant life on Earth. Perhaps fresher water existed at earlier times, but these acidic brines were present over 4000 million years ago.

The presence of methane in the Martian atmosphere at ppb levels has been reported by several observers. If confirmed, it might be of organic origin, although inorganic sources are possible. Methane can only have a short life in the atmosphere and its removal would in a few thousand years, deplete the very low level of oxygen present. The methane spectra are at the limits of resolution and non-unique, making the claim doubtful. Extraordinary claims require extraordinary proof.

However, if the presence of ancient life on another planet is eventually confirmed, it will have considerable philosophical interest. It will tell us that life may arise anywhere if the chemical conditions are right. It is of course a very long way from simple bacteria to the "Little Green Men" that so many people want to believe in. Furthermore, if life did arise on Mars, and died out, then life was clearly unable to influence the environment on Mars and so ensure survival by adapting the planet. This would constitute both a test of and a disproof of the Gaia hypothesis of James Lovelock, that the Earth is alive.

VENUS

Venus, rising in the morning or setting in the evening skies, is the most brilliant object in the sky, after the Sun and the Moon. This gleaming jewel has been admired since antiquity. Babylonian astrologers called it "the bright queen of the sky" in about 1600 BCE. Because it is an apparent twin of the Earth, it has always been of intense interest as the only Earth-mass planet in our planetary system. When Venus was found to have an atmosphere, it did not take much imagination to make it a hotter version of the Earth. This idea of a tropical version of the Earth seemed reasonable, as Venus was closer to the Sun. It was soon clothed with thick tropical forests and swamps, populated with various monsters. Dinosaur-like creatures were favorites of science fiction writers. The coal-forming swamps of the Carboniferous Period on the Earth, complete with giant dragonflies and exotic trees, provided other likely models for fevered minds.

Early observers thought that the planet was either spinning rapidly, similarly to the 24-hour period of the Earth, or perhaps only once a month. But a surprise was in store. When Earth-based radar in the 1960s penetrated the massive cloud cover, Venus was discovered to rotate very slowly, backwards. Unique among our planets, it takes 243 days to make one rotation on its axis, although the atmosphere of Venus rotates in about 4 days. Venus orbits the Sun in only 225 days, so that the day on Venus is longer than the year.

Unlike the Earth, Venus has no moon. The planet has a thick atmosphere, mostly of carbon dioxide. The atmospheric pressure is 93 times that of the Earth. There is only a trace of water vapor in the atmosphere, and none on the surface, where the temperature is 460°C (735 K), twice as hot as a kitchen oven and hot enough to melt lead. Whether there is water trapped deep in the interior remains an open question, but the evidence for a wet mantle is slight. It is likely that the planet is almost totally dry.

To complete this litany of differences, Venus has no detectable magnetic field. A navigator would be hard-pressed to find his way, as the stars would not be visible through the thick atmosphere either.

There is thus a certain irony that the surface of Venus was revealed in stunning clarity by the radar on the spacecraft named after Magellan, the celebrated explorer. The landscape that it has uncovered must inspire a sense of humility, since it is so different from that of the Earth. No hard-won models of geological evolution, worked out from long and diligent study of the surface of the Earth, are of any use on Venus. We have to start anew. This is a common tale throughout the solar system. Each newly explored planet and satellite has some variation from our knowledge gained by studying our own planet. Although the geological processes remain similar, they produce startlingly different results.

Venus is a little smaller in radius than the Earth, but has the same density, when allowance is made for the small size difference. So there is probably not any real difference in bulk composition. The apparently trivial difference in size turns out to be one of the crucial factors in making Venus different to the Earth. In contrast to the hopes of science fiction writers, Venus only looks like the Earth, in the same sense that the evil Mr Hyde resembled the good Dr Jekyll, in the novel by Robert Louis Stevenson (1850–1894).

The Earth's crust is divided into a thick, buoyant continental crust and a thin, dense crust of basalt that underlies the oceans. Like a baker who can only make one loaf, Venus seems to have produced a monotonous crust of basalt, perhaps 30 km thick. There is no sign of plate tectonics and nothing looking like the mid-ocean ridges or deep trenches that we see on the Earth. Thus there are remarkable secondary differences between the two planets, when one considers that they are close in density and size.

What are the causes of the differences? Many are accidental consequences of both planets forming from the assembly of rocky planetesimals in the inner solar system. Perhaps no big body hit Venus, so that it spins slowly. It may have no moon for the same reason.

Nothing struck the planet at the right angle to splash one off. The thick atmosphere remains because there is no ocean to absorb the CO_2 and nothing big enough hit the planet to remove it. The high surface temperature follows from this. The similarity in density to the Earth and the obvious presence of basaltic lavas imply a similar internal structure of a metal core and a rocky mantle. The absence of a magnetic field seems to be due to the slightly smaller size of Venus. It is the freezing of the Earth's inner core that is thought to drive the dynamo and so produce our magnetic field. The pressure deep inside Venus is just a little too low to make a solid inner core. Thus Venus and the Earth, although so close in size and uncompressed density, have evolved very differently. So significant differences between Earth-mass planets can arise from small changes in size.

Over a dozen large plates that jostle one another cover the surface of the Earth. Venus, in contrast, is a one-plate planet. If the water and the thin veneer of mud were removed, the basalt-covered floors of our oceans would look somewhat like the surface of Venus. However, it is clear that there are several major differences. There is no equivalent on Venus to the great mid-ocean ridges on Earth, where fresh lava comes to the surface, and from which the sea floor spreads away. Although there is local lateral movement on Venus, extensive spreading that is so characteristic of our oceanic crust does not occur. Nor is there any sign on the surface of Venus of the great deeps where the oceanic crust is pulled back down and recycled into the mantle. On Venus, there is nowhere for the lavas to go. Venus has choked itself with a thick crust of lava.

Unlike the Earth, the bone-dry rock at the surface of Venus is very strong. It can hold up steep slopes for millions of years. The strength of the crust on Venus is due to the absence of water. Like an armadillo, Venus has encased itself in a strong, dry, rigid shell of basalt. On the Earth, the mountains float like icebergs, supported by deep roots. On Venus, they just sit on the surface, which is strong enough to hold them up like the mythological Atlas, who carried the whole Earth on his shoulders. Some of these mountains are as high as those on Earth. The volcanic massif of Beta Regio is nearly 10 kilometers high. The major mountain range, the Maxwell Montes, rises 11 kilometers above the low rolling plains that cover most of Venus. These high-standing areas that might be mistaken for continents are folded basaltic volcanic crust.

There are three main regions. The oldest parts appear to be crumpled-up crust called tesserae. Most of the surface is covered with featureless rolling plains of basaltic lavas. These seem only a little younger than the tesserae. Other features are volcanoes and some circular structures a few hundred kilometers in diameter, called coronae. On the volcanic plains, there are thousands of small shield volcanoes, a few kilometers in diameter. Over 50,000 have been identified. They are just sitting on the surface, supported by the strong crust.

A few small "pancake" domes, about 20 kilometers across, appear to be composed of more viscous material, that has spread out like treacle. They may be like small domes of viscous granite on the Earth, but ours have rough jagged surfaces. Those on Venus appear smooth. The "pancakes" on Venus are isolated and not to be confused with the broad expanses of granite on our familiar continents. The production of great masses of granite, as we see on our continental crust, does not occur on Venus. This rock, so familiar to us from city buildings and pavements, is uncommon in the rest of the solar system.

Is there a decipherable geological sequence or history on the planet? Probably not. In spite of the presence of many craters, they are not abundant enough to establish a relative sequence among the tesserae, plains, volcanoes or coronae. Attempts at constructing a geological chronology fall back on resemblances between similar looking rocks or terrains, a hazardous procedure shunned by experienced geologists. Although this twin of the Earth has a similar heat production, Venus has had a totally different geological history. So there is no agreed geological timescale for the planet.

The craters formed by meteorite impact on the surface of Venus are surprisingly fresh. Only rarely are they obscured or flooded by lava flows. No craters less than 5 kilometers in diameter have been detected on Venus and there are hardly any craters with diameters less than 30 kilometers. This is a tribute to the thick atmosphere of Venus. The smaller meteorites explode or are burnt up due to friction as the incoming body plows through the dense clouds. Our own atmosphere, 50 times less dense, is a much weaker shield. However, it protects us from the smaller bits and pieces that we see as "shooting stars" on clear nights in the country.

There are also numerous dark, smooth patches or "splotches" on the plains of Venus, extending over many kilometers. These are apparently caused by shock waves that blast and scour the surface. These result from meteorites that are too small to penetrate through to the surface and which explode high in the atmosphere. The Earth had a similar experience in 1908, when a meteorite about 60 meters in size exploded 5 kilometers above the Tunguska River in Siberia and blew down the surrounding forest.

It seems appropriate that the planet named for the goddess of love should have a young face. The age of the surface of Venus can be worked out from the number of craters that have formed on it over time. We know the rate of impacts from our study of the craters on the dated surfaces on the Moon. Based on this, the surface of Venus appears to be relatively young. This is a very strange finding. On Earth, we are missing only the first 10% of our history in the rock record, but on Venus, there is no record of what happened in the first 85% of its history. There are no old surfaces covered with craters, such as we see on Mars, Mercury or the Moon.

The present surface of Venus is about 750 million years old. At that time older surfaces were covered by lava. This was not however a catastrophic Noah-like flood of lava, but it probably took 100 million years to recoat the planet. Apparently exhausted by this great outpouring, geological activity on Venus hasn't managed to produce more than a trickle of lava since then. The presence of a few hot spots suggests that there may still be a few active volcanoes.

Although Venus had a similar production of heat to the Earth, our planet, in contrast, has rather carefully conserved its energy by recycling its basaltic oceanic crust back down into the mantle, using this process along the way to make our useful continental crust. Venus, in contrast, may have undergone vast outpourings of lava every few hundred million years. Even our own planet manages to produce occasional floods of lava, such as the Columbia River Basalts or Deccan Traps, but these are fortunately on a much smaller scale than on Venus.

One sobering aspect of the craters on Venus is that there are so many of them on a surface that is young by geological standards. Venus is close to the Earth in size and location so that our planet presents a similar target to this bombardment. This provides us with the chilling information that a large impact crater must also have formed on the Earth about every half million years. Most of this record has been erased by erosion.

Plate tectonics on Venus?

A basic question relevant here is why is there such a marked difference between the crustal development of the twin planets, Earth and Venus? They are so alike in size, mass and bulk density that they must be close in bulk composition. The abundance of the heat-producing elements, K, U and Th, is likely to be similar on Earth and Venus, an observation that merely exacerbates the problem.

The difference in their subsequent evolution is a consequence of the low water content of Venus compared with the Earth. This leads to a strong coupling between the dry crust and mantle. So unlike the Earth, there is no weak water-rich zone at 100 km depth on which plates might slide. On the Earth, the sinking sea floor crust converts to a denser phase at higher pressures. This sinking dense slab pulls plates down into the mantle. This process accounts for about 90% of the driving force for plate tectonics. But the smaller size of Venus means that pressures to effect this change are too deep to induce plates to sink.

All of these outcomes conspire on Venus and are the causes of the absence of any evidence for the operation of plate tectonics on this dry planet. So the contrast between the tectonics of Venus and the Earth is mainly due to the difference in water contents of the two planets, exacerbated by a slight difference in size. Trivial differences between Earth-mass planets affect habitability in significant and unpredictable ways.

Water on Venus

The source and history of water in Venus continues to be debated. The surface temperature of Venus is far above the boiling point of water and the atmosphere contains only a trace of water vapor. The present water budget on Venus is five orders of magnitude less than that of the water on Earth. But our own planet is already depleted relative to the primitive solar nebula abundance of water, by a factor of several thousand.

Did Venus ever possess more water? Were there early oceans? Because the Earth and Venus are commonly thought of as twin planets, it is natural to think of an early Venus covered with an ocean that vanished, usually ascribed to overheating by a runaway greenhouse effect. This is a current cause for alarm, that something similar will happen to the Earth if we put too much carbon dioxide into the atmosphere by burning fossil fuels.

There are various models for runaway greenhouses on Venus. Most invoke increasing temperature that caused water to evaporate, leading in turn to higher temperatures, more water vapor in the atmosphere, even higher temperatures and so on until the present hot, dry surface is achieved. In this concept, water dissociated into hydrogen and oxygen in the atmosphere. The leftover oxygen reacted with the surface rocks, leading to the present dry state of Venus. The lighter hydrogen escaped more easily than its heavier isotope deuterium, leading to a D/H ratio for Venus that is about 150 times greater than that of the Earth. This has often been construed as evidence for an initially wet or damp Venus and for a runaway greenhouse that evaporated an early ocean.

There are some problems with this scenario. Among them, oxygen needs removing. The usual solution is to use it to oxidize the surface. Such a process would require oxidation of a vast mass of material and might seem difficult to achieve on a planet encased in a crust of basalt, already mostly oxidized. The high D/H ratio in the atmosphere might equally well be due to later additions from comets from the Oort Cloud that have high D/H ratios. Perhaps all the water on Venus came from such a source.

But underlying all such explanations is the assumption that early Venus was "just like the Earth", complete with an ocean that had later vanished. But if the supply of water to the dry, rocky planets in the inner nebula was a random, "hit or miss" affair, there are no good reasons to suppose that Venus ever had much water to begin with.

As the study of Venus and the Earth shows, similarity does not mean identity. As we search for terrestrial-like planets elsewhere, we need to find out the reasons for these differences and the conditions that allow these diverse bodies, or Mercury and Mars for that matter, to form at all. Just as geology arose in the nineteenth century, now the study of planets represents a new area in scientific enquiry.

6 Earth and Moon

> "Ultimately, in the future of exoplanets, we would like an image of an Earth twin as beautiful as the Apollo images of Earth" [1].

Our planet is the very model of a habitable planet. Accompanied by its silvery Moon, the Earth forms our only current example of habitability. For this reason, these currently unique objects require a separate section, to consider why this might be so and how such a welcome situation came about. But the random processes involved during formation and the level of complexity exposed by subsequent geological and biological evolution give one pause in seeking clones among the exoplanets.

THE EARTH

John Updike remarked that "this planet is exceptional; clearly Venus or Jupiter wouldn't be agreeable to us" [2]. How did it manage to achieve this status? Can one deduce any general principles from a unique planet accompanied by a unique satellite? A good example has been the difficulty in recognizing on the Earth that the formation of craters by the impact of asteroids, comets and meteorites is an important planetary process. Erosion has removed most of the evidence. It is only within living memory that professional geologists have come to accept that the craters were not caused by internal eruptions. Our experience with the geology and geochemistry of the Moon, with its subtle but crucial distinctions from our hard-won experience on the Earth, should also remind us of the hazards of trying to extrapolate from unique terrestrial conditions.

The great philosophical contribution made by the study of the Earth was to establish the immensity of deep time which is illustrated by almost any geological feature. The 6000 years that Archbishop Ussher calculated have been replaced with an age for the Earth of 4.5 billion years, which is in itself only a fraction of the age of the rest of the universe.

So what does one say about this planet, unique even by the standards of our planetary system? We know so much detail about the Earth that any attempt to summarize this knowledge risks reducing the discussion to a trivial level or to a recitation of truisms. Here I discuss some salient points.

FORMATION

We have a reliable date of 4567 million years for T_{zero}, the formation of the solar system. When did the Earth form? Current models need about 100 million years to form the Earth and Venus from the hierarchy of dry rocky planetesimals. These had formed very early, along with Mars, Vesta and Mercury. Between half and three quarters of the Earth was put together from bodies the size of the Moon or larger. Clearly the random population of bodies left over in the inner reaches of the solar nebula decided the size and composition of the Earth and Venus.

The structure, composition and subsequent evolution of the Earth were determined by these chance events. A different location and population of colliding bodies might have resulted in something like Venus, or some entirely distinct planet.

The formation of the Moon was one of the last great collisions. Next, the molten Moon cooled and formed a crust. This is the next definite event that we can date. Thus a younger limit for the age of the Earth is given by the crystallization of the lunar crust at around 4460 million years ago. High temperatures that result from these colossal collisions are a consequence, so that total melting of the Earth seems unavoidable during the collision that formed the Moon. Molten metal, mainly iron, fell into the center, like iron in a blast furnace, so that the metal core of the Earth formed quickly.

The inner part of our metallic core has now cooled and is solid, but the outer part is still molten, something we understand from studying the passage of earthquake waves. The rocky mantle cooled and became solid very quickly, but just how this happened remains a mystery.

We understand very well how this occurred on the Moon, but the sheer size of the Earth has defeated the attempts of modelers to understand what happened here. There is no sign of the segregation of minerals that occurs in small pools of melted rock on the Earth. A similar process that formed different zones of minerals also occurred on a larger scale during the cooling and crystallization of the Moon from its initially molten state. Probably the Earth cooled very quickly, before the crystals had a chance to separate.

The minerals that form the rocky mantle of the Earth are dominated by the common elements such as iron, magnesium, aluminum, silicon and sulfur. The relative sizes of the atoms determine whether they can fit into the growing crystals or are pushed aside, like square pegs trying to fit into round holes. Most of the other chemical elements are present in trace amounts, so they have to scramble around trying to find the minerals with the appropriate size cavities that will take them in. Many elements are too large, too small or have the wrong valency to fit into the common minerals, such as olivine, pyroxene or feldspar, and so rare elements are mostly concentrated in residual fluids, which often contain high concentrations. By such processes, trace elements like the rare earths are sometimes concentrated enough to form ore deposits.

How does the bulk composition of the Earth compare with that of the other planets? The abundances of the chemical elements in the primitive meteorites such as the CI class of carbonaceous chondrites give us the best estimate of the composition of the rocky component of the solar nebula. One might have thought that the composition of the rocky planets like the Earth would match this. Many theories of planetary formation have often supposed so. However, the terrestrial planets have lost not only the gas and ice in the original nebula, but are also depleted in elements such as lead, sodium and potassium that are volatile at temperatures less than 1000 K (Figure 4, Figure 5). Venus has about the same bulk composition and a similar budget of heat-producing elements, as far as we can judge. Mars has a lower bulk density and a higher content of volatile elements. Mercury has much more iron compared to silicate. So the composition of a rocky planet depends on prior events in the nebula, a chance assembly of planetesimals and the history of collisions.

CRUST, MANTLE AND CORE

The composition of the upper crust of the Earth is quite well known. After all we live on it and geologists have been hammering, probing and drilling it for 300 years. The crust under the oceans, which covers about three quarters of the surface of the Earth, is mainly basalt, about 5 or 6 kilometers thick, erupted as lava from the mid-ocean ridges and coated with a veneer of mud that comes from the erosion of the continents.

The composition of the rocky upper mantle, down to about 200 kilometers, is also relatively well understood, as we have samples from there brought up by volcanoes. Deeper down, our knowledge is less certain, although we know from studying the passage of earthquake waves that the mantle is rocky down to the core. After several billion years of producing lavas that are recycled back into it, our mantle resembles a "marble cake" or "plum pudding", although the plums are rare enough to present problems for Jack Horner.

The core, which begins at 2900 kilometers below the surface, is mostly metallic iron, alloyed with about 10% of nickel, some sulfur and other metal-loving elements such as gold and platinum. However, the rocky mantle of the Earth has more nickel, platinum, iridium and similar metallic elements than expected. By rights these elements should all be in the core along with the iron and we would have no platinum wedding rings. So the current model is that these metallic elements were added from meteorites sprinkled in late in the formation of the Earth. Such late veneers needed to be thoroughly mixed into the rocky mantle, which was probably molten at the time. It was also at this time that, along with these elements, the Earth received much of its complement of the very volatile elements.

The most important addition to what was a dry planet was the arrival of 500 ppm of water, courtesy of the addition of a few (possibly only one) icy planetesimals or comets from near Jupiter. This random event had momentous consequences. Life was able to get started, oceans to form, plate tectonics to arise, followed by the formation of the continental crust and eventually *Homo sapiens*, all the outgrowth of the chance arrival of a splash of water.

Geology

> "Geology is about what happened – not what should have happened" [3].

The study of the Earth, or geology, is frequently thought to be an easy subject, only a little more difficult than geography, and so it is popular with those students anxious to avoid the rigors of physics or calculus. But its apparent simplicity is deceptive. Contrary to popular wisdom, geology is one of the more difficult of the sciences.

This is readily demonstrated by considering the historical development of the various sciences. Thus classical physics was well established by Newton, with the publication of the *Principia* in 1687. Darwin set biology upon the right track in 1859 when he published *The Origin of Species*. The underlying basis of chemistry became understood with the formulation of the Periodic Table of the Elements by Dmitri Ivanovich Mendeleev in 1869. Ernest Rutherford established the fundamental nature of atoms a century ago in 1911. Even the origin of the chemical elements themselves was understood following the work of the Burbidges, Willy Fowler, Fred Hoyle and independently by Al Cameron in 1956.

But it was only as late as 1963, three centuries after Newton's physical insights, that Fred Vine and Drum Matthews hit upon the fundamental process of plate tectonics. Then geologists finally understood what was going on under their feet. This mechanism explained the architecture of the surface of the Earth which had been painfully established in the previous 150 years following the pioneering works of James Hutton, Georges Cuvier, William Smith and Charles Lyell (Figure 16).

FIGURE 16 Siccar Point, Scotland

The Earth records its history. This immensity of geological time is well shown in this picture of the famous unconformity at Siccar Point, near St Abb's Head on the east coast of Scotland, north of Berwick on Tweed. The slightly inclined rocks on top are the Old Red Sandstone that have been uplifted and tilted about 15°. The vertical strata underneath them had long before been sands and muds, laid down on the sea floor in horizontal beds, before being tilted through 90°, raised above sea level and planed off by erosion. They next sank below sea level and formed the base on which the sands of the Old Red Sandstone were deposited in the Devonian Period (in popular language, the "age of the armored fishes"). The term "unconformity" refers to the boundary between these two sets of strata. The underlying vertical strata are Silurian in age and this outcrop records the passage of about 50 million years. Although this is inconceivably long on human timescales, it constitutes less than 1% of geological time. Such examples abound in the geological record. John Playfair (1748–1819) wrote on visiting this outcrop in about 1790, "We felt ourselves . . . carried back to the time when . . . the sandstone before us was only beginning to be deposited, in the shape of sand and mud, from the waters of an . . . ocean . . . An epoch still more remote presented itself, when even the most ancient of these rocks, instead of standing upright in vertical beds, lay in horizontal planes at the bottom of the sea . . . revolutions still more remote appeared in the distance of this extraordinary perspective. The mind seemed to grow giddy by looking so far into the abyss of time", from John Playfair, *Illustrations of the Huttonian Theory of the Earth*, 1802. Facsimile reprint by G. W. White, University of Illinois Press, 1956, xiii–xiv (photo credit Dave Souza, Wikipedia). See also color plates section.

Their work, along with that of innumerable other geologists, led to the establishment of the geological timescale, based mostly on the fossil evidence from the last 500 million years. This triumph of Victorian science has proven robust and the well-known subdivision of time into such Periods as the Permian and Cretaceous has survived (see Appendices).

But it has been difficult to find geological laws or generalizations of general applicability such as the Hertzsprung–Russell diagram or the Periodic Table of the Elements that enabled the rapid development of astronomy and chemistry. Such problems are responsible for the lengthy development of geology and of the continual reappearance of bizarre theories from amateur and professional geologists alike, to account for geological phenomena. These are some of the reasons that the development of the geological sciences lagged behind that of most other sciences.

THE GEOLOGICAL EVOLUTION OF THE EARTH

> "Planetary history, like planetary formation, is dominated by stochastic and unpredictable events" [4].

Fortunately the Earth records much of what has happened. Several strands can be discerned. The first is the organization of the gross structure of the planet into a metallic core and rocky mantle, a consequence of early melting. This was followed on this planet by the development of a crust, resulting in the generation of plate tectonics and eventually in the creation of the continental crust. But there are difficulties in trying to discover some general patterns for the formation of crusts on rocky planets. Crustal formation on the Earth or Moon does not necessarily provide us with a model for the development of crusts on Mercury, Vesta, Mars or Venus [4].

Both the physical aspects of geological evolution and of life were intertwined, with life responding to the geological changes imposed on the environment. But many of these are contingent, accidental events that include such upheavals as asteroid strikes, massive floods of lava, release of trapped methane, "snowball" Earths and the rise of an oxygen atmosphere, that are superimposed on the slower processes of geology. All have had dramatic effects on the evolution of life [5].

Each terrestrial planet is unique. Although features such as sand dunes arise on the Earth, Venus, Mars and Titan due to the action of wind on grains, these have distinct origins on each body. Indeed, the sequence of geological events on the Earth itself has little predictive power. If one had visited the Earth during the Permian, one would not have foreseen the world of the Triassic with its completely different fauna. To a visitor in the warm Cretaceous, it would have been difficult to imagine the cooling throughout the Tertiary or the onset of the ice ages. Nor could a visitor to Venus a billion years ago have foreseen the total resurfacing of the planet that was to occur shortly thereafter.

But even less predictable was the catastrophe that would end the benign Cretaceous, remove the giant reptiles and lead to the dominance of mammals. That event, which has now resulted in the Earth being overrun by one species, was one consequence of the collision of the Earth with a 10-km-diameter asteroid, at the end of the Cretaceous.

The Hadean environment

A geological timescale, listing the various Eons, Eras and Periods into which geologists have divided both time and the rock record, may be found in the Appendices. A basic question is what happened in the gap of about 500 million years after the formation of the Earth and before that of the oldest rocks that we recognize? This absence of an unequivocal rock record has not prevented – indeed, has encouraged – speculation about the nature of the Hadean Earth.

The nature of the early crust remains enigmatic. It was not made of feldspar, as on the Moon, nor was it granitic, but was likely basaltic in composition. Small amounts of granitic rocks differentiated from this basaltic crust and emerged as islands from an early ocean. But substantial masses of granite, as in our present continents, were not present. Erosion of these would have contributed swarms of resistant minerals such as zircons to younger sedimentary rocks. But zircons or other resistant grains eroded from rocks older than 3900 million years are rare, consistent with derivation from a few scattered outcrops.

Although the Hadean is often viewed as a Dantean version of Hell, accompanied by a continuing bombardment, impacts following the initial accretion were probably haphazard occurrences until the Late Heavy Bombardment occurred around 4000 million years ago.

If one traveled back into the Hadean, one might view from a time capsule a broad ocean with rare islands of granite, of which the zircons are the only surviving remnants. A submersible would encounter basalt just as on our present sea floor. Rarely the impact of an asteroid would disrupt the tranquil scene.

The Late Heavy Bombardment

A period of heavy bombardment ended the Hadean, complicating all interpretations of the early Earth. Between 4100 and 3850 million years ago, the Moon and most likely the Earth, were struck by large numbers of asteroids or comets. The cause was probably a consequence of movements among the giant planets. These precipitated a shower of bodies into the inner reaches of our solar system, as elegantly described by the Nice Model.

We know the dates of this event rather precisely from our study of the samples from the Moon. Models suggest that over 200 great basins, bigger than France, likely also formed on the Earth during this period. However, extensive searches for traces of this bombardment in Early Archean rocks have failed to find evidence of this catastrophe. Neither shocked minerals nor a geochemical signature (such as excess iridium) characteristic of meteoritic impacts have been identified. The simplest explanation is that the oldest surviving rocks on the Earth postdate the cataclysm that ended at 3850 million years ago, so that the record is simply missing.

Archean times

On Earth, the Hadean was followed by the Archean, where for the first time we have rocks that geologists can interpret. The enigmatic

Archean covers a crucial 1500 Myr of Earth history, from the first traces of continental rocks to the formation of major continents in the Late Archean. Basaltic crusts formed early but they were thick and buoyant, a consequence of a hotter early Earth. But they began melting at their base, producing granitic-looking rocks. These had more sodium than our familiar granites, which contain more potassium. Together with basalt, these sodium-rich granites dominate the Archean crust.

By the end of the Archean, about two thirds of our continental crust was present and our familiar cycle of plate tectonics began. Thus the last 500 million years of the Archean is a crucial period in Earth history. It reflects a fundamental change in crustal evolution, which is linked to the long-term cooling of the Earth.

The modern plate tectonic regime began to operate at around 3000 million years ago, when 40% of Earth history had already occurred. By that time sinking of the cooler oceanic crust became possible. This change was related to the long-term reduction in heat production and the formation of thinner oceanic crust. Convection currents deep within the Earth ultimately drive plate tectonics.

The internal heat engine of the Earth gets energy from two sources, both of which decay with time. One is the gravitational heat resulting from the assembly of the planets by large planetesimals. The other comes from the decay of radioactive elements, notably potassium, uranium and thorium. Heat production from this source was about five times greater in the Hadean, but had fallen to about twice its present value by the late Archean and has been falling off ever since. Now it contributes about half the heat driving the geological processes in the Earth, the rest coming from the residual heat of formation. This energy drives the slow cycling of the mantle and plate tectonics.

Impacts had not ceased with the Late Heavy Bombardment. During the Archean, some 10-kilometer bodies were still wandering about, occasionally colliding with the Earth. This is sometimes referred to as the "Late Late Heavy Bombardment". The record of craters in the Archean is missing, but several beds of spherules, up to 30 cm thick, record fallout from these large collisions. These spherules apparently formed in the atmosphere from vaporized silicates. Some have traces of the element iridium from the impactor, although nothing seems to match known meteorites. Possibly there were 10 or so of these impacts, around the size of the Cretaceous–Tertiary (K-T) impact. Although they are often postulated to have produced dramatic geophysical consequences affecting subsequent geological developments, the scale of even the largest impacts is too small to cause more than local melting.

But something even more significant happened in the Archean. Life originated. There are very controversial suggestions that life had become established by about 3850 million years ago. The presence of sedimentary rocks at that distant epoch tells us that water was present on the surface of the Earth. Ancient sedimentary rocks in Greenland that formed in that remote time bear a highly debatable signature of microbial life in the ratios of their carbon isotopes.

Reports continue to appear of bacterial fossils in the Early Archean, but remain equally controversial. However, life in the form of sulfur-using bacteria seems to have been clearly in existence by the Mid-Archean, 3400 million years ago, nearly 1 billion years after the Earth formed. Stromatolites that formed from sediment trapped by mats of cyanobacteria are found that are a little younger in age. Curiously, similar stromatolites are forming at present not far away in Shark Bay, on the west coast of Australia, an outstanding tribute to the ability of life to survive on this planet for over 3 billion years. This also tells us that conditions on the surface of the Earth have not changed drastically in the region over that immense length of time.

The origin of life thus seems to have occurred sometime between 0.5 and 1 billion years after the planet formed. But such life as existed on Earth was restricted to simple single cell organisms, called procaryotes, for immense periods of time. What does this tell us about the probability of life appearing on suitable exoplanets? Probably not much, except that there are many unrelated and contingent steps between the origin of life and the development of an intelligent species.

The Proterozoic

Abrupt changes in crustal evolution and life occurred at around the end of the Archean. The next great discernible span of Earth history is labeled the Proterozoic Eon, which occupies a stupendous 2 billion years between the end of the Archean and the first appearance of fossils with shells in the Cambrian. As the heat production of the Earth declined, convection within the mantle slowed, the oceanic crust became cooler, thinner and less buoyant. The spreading sea floor encountered the great masses of thick low-density rocks, obstructions resembling icebergs, and sank back into the mantle. As it sank, increasing pressure transformed the minerals in the basalt into denser phases. This increase in density dragged the sinking crust further into the mantle. So began the saga of plate tectonics that has led to many consequences. Although we place these zones of time into neat pigeonholes, there was much overlap and these events spread over several hundred million years from the late Archean through to the Early Proterozoic.

Early in the Proterozoic, several other major events occurred. The great masses of continental rocks exposed large rock surfaces to the atmosphere. Carbon dioxide combined with water in the atmosphere, producing acid rain that weathered the rocks and transported elements such as calcium and magnesium to the oceans, resulting in depletion of the carbon dioxide in the atmosphere.

About 2000 million years ago, life that had hitherto consisted of single-celled bacteria and Archea, the so-called procaryotes, developed complex cells with nuclei, the so-called eucaryotes that form the basis for later complex life. As Preston Cloud (1912–1991) commented, "the appearance of the eucaryotic cell was a Proterozoic triumph – the main event of biological evolution after the origin of life itself" [6].

Equally significantly, photosynthetic bacteria capable of producing oxygen became abundant. These seem to have evolved at around 2700 million years ago. Initially, the oxygen produced oxidized the large concentrations of reduced iron in the oceans, forming the banded iron formations that today supply us with much of our iron ore. The arrival of free oxygen also began to remove the methane from the atmosphere that had dominated the Archean and had perhaps kept it warm during the period of the faint early Sun. The CO_2 was already being drawn-down by weathering of the extensive continents. By about 2400 million years ago, sufficient oxygen had built up to dominate the atmosphere.

The arrival of oxygen is sometimes called "The Great Oxygenation Event", the "Great Oxidation" or "The Oxygen Catastrophe". Who was affected by the catastrophe depends on ones point of view but it certainly was dramatic for the methane-producing bacteria.

The removal of methane and CO_2 from the atmosphere is commonly thought to be the cause of the earliest widespread glaciation. This is labeled the Huronian glaciation, which occurred 2400 million years ago, named after the excellent record on the shores of Lake Huron. Ironically, this is one of the great lakes resulting from the last ice age. The Huronian Glaciation was particularly widespread and long lasting. Indeed the Earth may have frozen over, creating a "Snowball Earth". All of these events occurred close to the Archean–Proterozoic boundary, formally marked at 2500 million years ago.

It marks a great change in the geological evolution of the Earth. A consequence of plate tectonics was that, as the crust under the oceans disappeared downwards, it encountered not only high pressures but also high temperatures. These drove off water and the more volatile elements that rose into the mantle above the subsiding crust. This flux of volatile material induced melting in the mantle. This resulted in the eruption of the great andesite volcanoes, explosive on account of their high water content, which added new material to the growing continents. Large amounts of the radioactive elements were transferred upwards from the mantle into the growing continental crust. This heat source, perhaps aided by heat from the mantle, induced additional melting within the crust itself. These melts rose

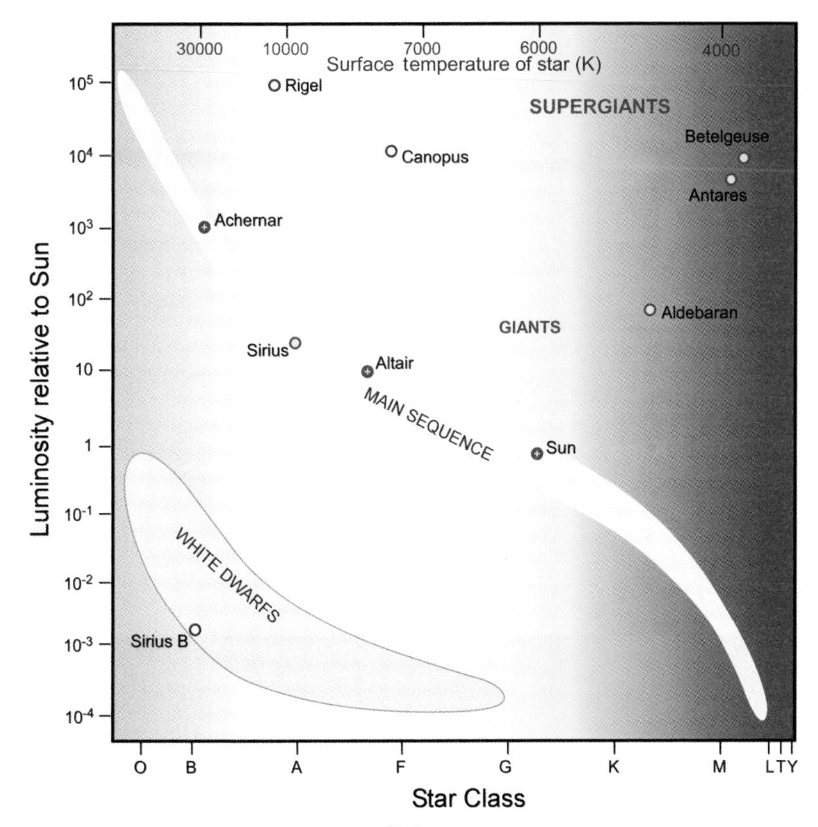

PLATE I Hertzsprung–Russell diagram

The famous Hertzsprung–Russell diagram plots the surface temperature of a star (or class of star) against the luminosity of the star relative to the Sun on a logarithmic scale. The star classes include L, T and Y classes (brown dwarfs) that lurk in the bottom-right corner, as well as the classical OBAFGKM classes. Some well-known stars are identified.

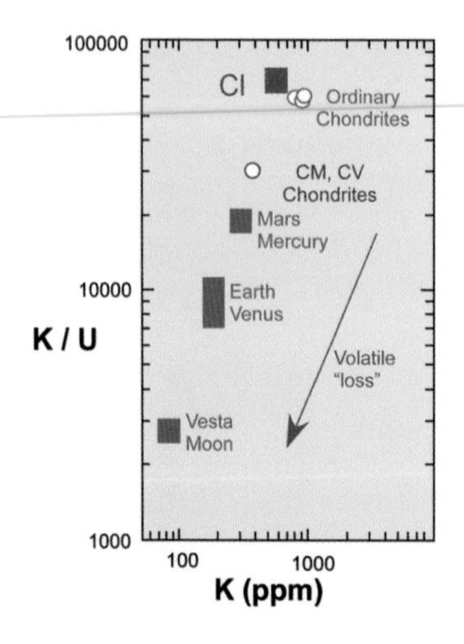

PLATE 2 Volatile element depletion

The widespread depletion of volatile elements in the inner solar system, displayed here as the abundance of potassium (K), a "volatile" element relative to uranium (U), a "refractory" element. CI gives the composition of the "rocky" fraction of the Sun and of the primitive solar nebula, as explained in the text. The chondrites are various classes of stony meteorites (courtesy Scott McLennan).

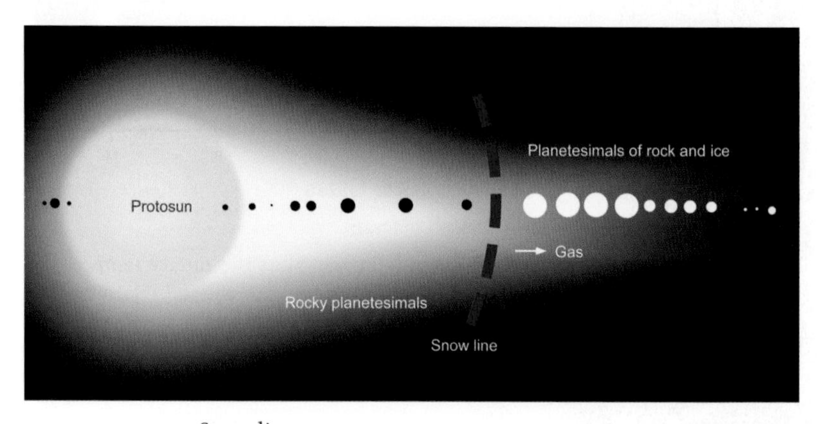

PLATE 3 Snow line

As the early Sun turned on its nuclear furnace, strong stellar winds swept water and other volatile material out to between 3 and 5 AU, where the water condensed as ice and piled up in a snow line. This increased the density of the nebula at this location and so enabled icy cores about ten times the mass of the Earth to grow quickly. Two of these large cores then captured some of the gas that was also being driven away from the violent early Sun and became the gas giants, Jupiter and Saturn. The cores of Uranus and Neptune, farther out, only managed to catch a little gas. The terrestrial planets formed from the dry rock rubble left sunwards of the snow line. (Adapted from a slide courtesy of John Wood.)

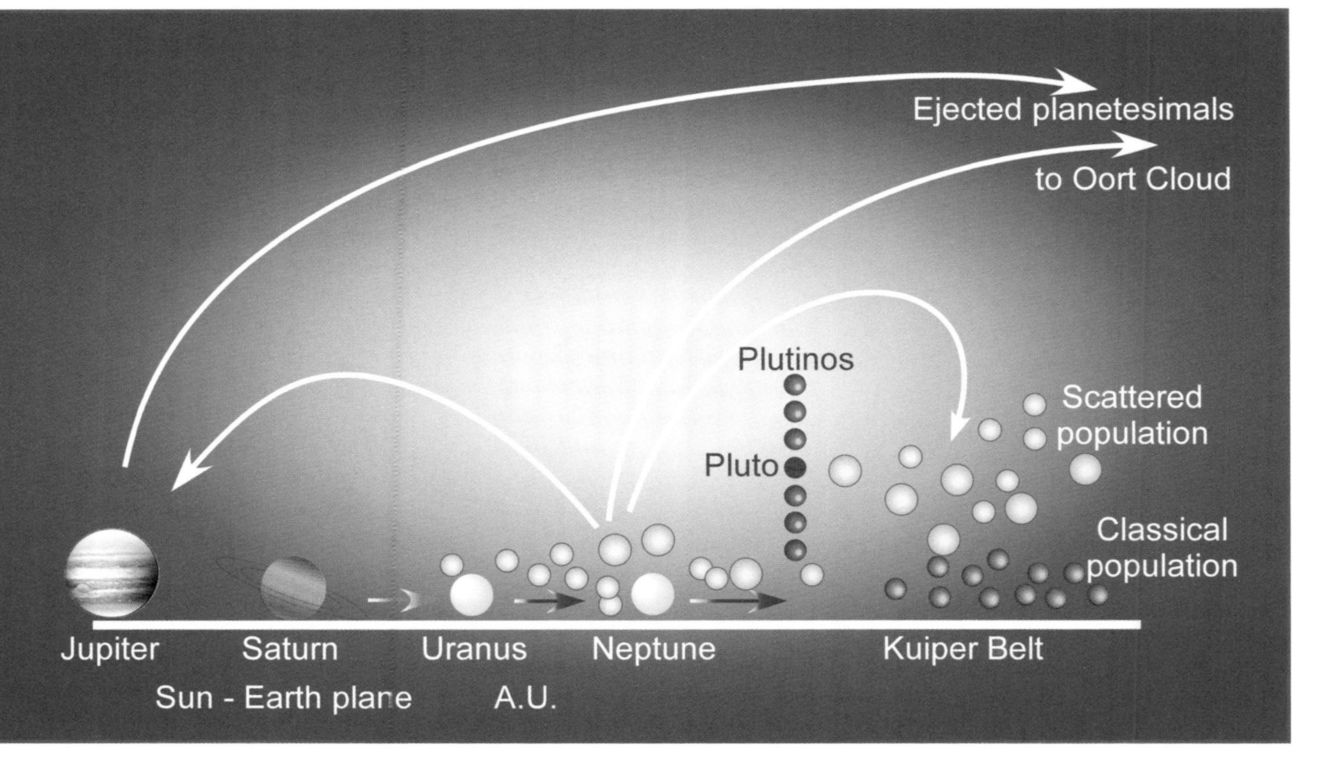

PLATE 4 Nice Model

A representation, not to scale, of one stage of the migration of the giant planets according to the Nice Model. Saturn, Uranus and Neptune are moving outwards, scattering the residual icy planetesimals out into the Kuiper Belt and the Oort Cloud. Some, including Pluto and the Plutinos have been captured into resonant orbits with Neptune (adapted from Morbidelli, A. and Levison, H. *Nature* Vol. **422**, p. 31, 2003.)

PLATE 5 Burns formation, Mars

Planets record their history. Ancient sedimentary rocks, about 3.7 billion years old, at Meridiani Planum on Mars, photographed from the Opportunity Rover. The Burns formation is composed of cemented sandstones. They were originally sand dunes in a desert. When groundwater reached the surface briefly, temporary small inter-dune playa lakes formed. See text. (Courtesy Scott McLennan.)

PLATE 6 Siccar Point, Scotland

The Earth records its history. This immensity of geological time is well shown in this picture of the famous unconformity at Siccar Point, near St Abb's Head on the east coast of Scotland, north of Berwick on Tweed. The slightly inclined rocks on top are the Old Red Sandstone that have been uplifted and tilted about 15°. The vertical strata underneath them had long before been sands and muds, laid down on the sea floor in horizontal beds, before being tilted through 90°, raised above sea level and planed off by erosion. They next sank below sea level and formed the base on which the sands of the Old Red Sandstone were deposited in the Devonian Period (in popular language, the "age of the armored fishes"). The term "unconformity" refers to the boundary between these two sets of strata. The underlying vertical strata are Silurian in age and this outcrop records the passage of about 50 million years. Although this is inconceivably long on human timescales, it constitutes less than 1% of geological time. Such examples abound in the geological record. John Playfair (1748–1819) wrote on visiting this outcrop in about 1790, "We felt ourselves . . . carried back to the time when . . . the sandstone before us was only beginning to be deposited, in the shape of sand and mud, from the waters of an . . . ocean . . . An epoch still more remote presented itself, when even the most ancient of these rocks, instead of standing upright in vertical beds, lay in horizontal planes at the bottom of the sea . . . revolutions still more remote appeared in the distance of this extraordinary perspective. The mind seemed to grow giddy by looking so far into the abyss of time", from John Playfair, *Illustrations of the Huttonian Theory of the Earth*, 1802. Facsimile reprint by G. W. White, University of Illinois Press, 1956, xiii–xiv (photo credit Dave Souza, Wikipedia).

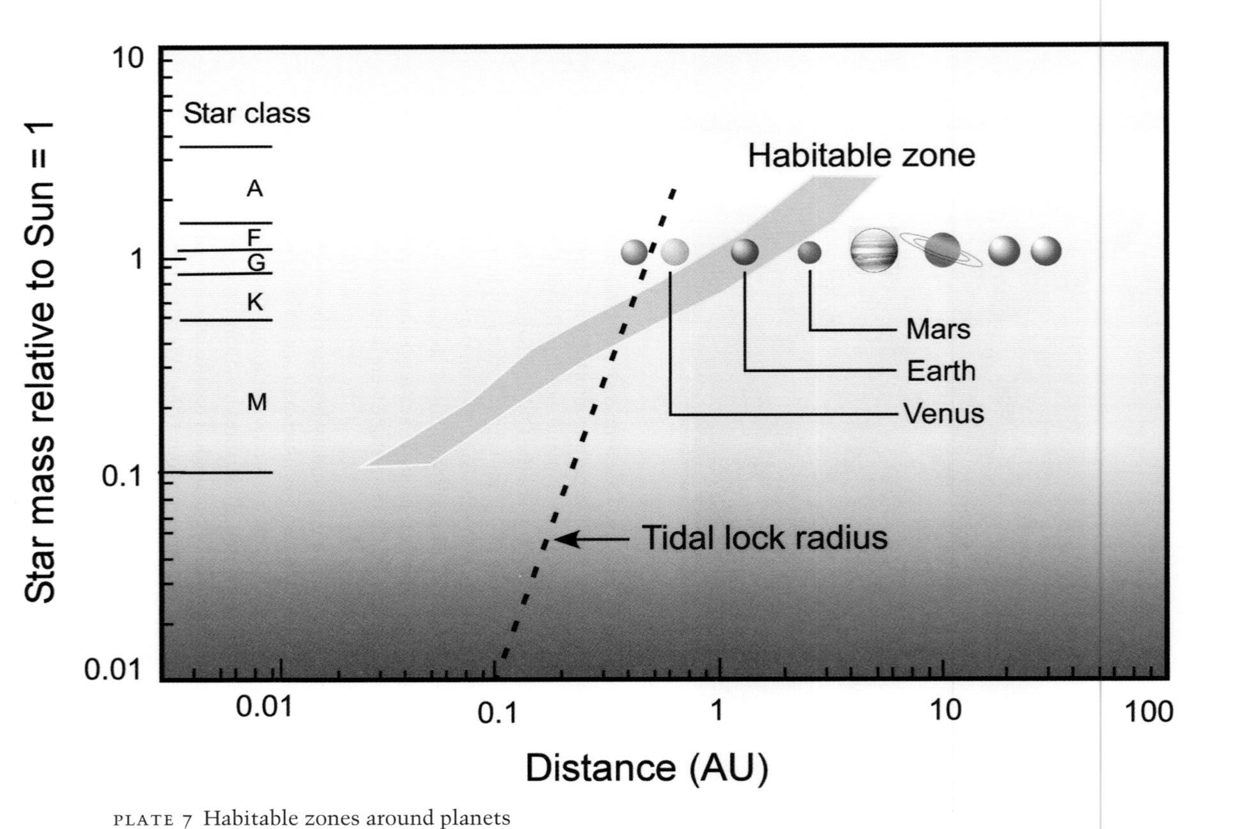

PLATE 7 Habitable zones around planets

The "habitable zone" in which liquid water is stable on a planetary surface, plotted against the classes of stars, with our planets shown opposite the G class (not to scale). As the mass of the parent star decreases, the habitable zone moves closer to the star. Planets in such locations may be tidally locked into synchronous rotation, always presenting one face to the star in an eternal gaze.

PLATE 8 Former civilizations

Traces of former sophistication. The temple of Kukulcan at Chichen Izta in Yucatan, Mexico built in 445 BCE, and used until 1204 CE The temple (El Castillo) accurately predicts the Spring (March 20 or 21) and Fall (September 21 or 22) equinoxes, a tribute to Mayan science. The base width is 55 meters (180 feet) and the structure rises 24 meters (78 feet) surmounted by a 6 meter (20 feet) temple. The site of the K-T impact was also in Yucatan. The sacrificial well at this site is one of the sinkholes (cenotes) that mark the edge of the impact crater and which are probably related to subsidence along the crater rim.

PLATE 9 Deep Time

Beachy Head, a cliff made entirely of chalk, on the south coast of England, rises 162 meters or about 520 feet. The chalk was deposited by a rain of coccolith (nanoplankton) fragments a few microns in size, on to the Cretaceous sea floor that occupied much of what is now Western Europe. The overall accumulation rate was at a rate of 1 millimeter per century, so that the chalk exposed in this cliff represents the passage of over 16 million years.

into the upper crust, producing our familiar granites that now dominate our continental crust.

The initial form of the continental masses is lost in the mists of time. But by late in the Proterozoic, the continental masses formed a single block, called Rodinia. Because of convection currents in the mantle it began to break up. The separate continental fragments changed the oceanic currents, which precipitated the great glaciation called the "Snowball Earth". Later movements assembled other ancient supercontinents, including Pangea, Gondwana and Laurasia.

Major developments in life appeared, perhaps all the result of these geological transitions, including major glaciations that may have led to a "Snowball Earth" [7]. Large complex multi-celled, but soft-bodied animals begin to appear in rocks from the Late Proterozoic. We know about these easily destroyed creatures, as they were buried suddenly under showers of volcanic ash or mudflows on the sea floor. The type example is in South Australia, from where the name Ediacaran comes. This term is given to the period of time 575–542 million years ago. Many bizarre forms appear that represent failed experiments in evolution. But this strange fauna vanished abruptly before the onset of the Cambrian fauna. Did it vanish as the result of a mass extinction or were they just eaten by more advanced forms?

Collisions with asteroids extended into the Proterozoic and beyond. Giant impacts formed the Vredefort (2023 million years ago) and Sudbury (1850 million years ago) structures. The celebrated dinosaur-removing collision at the Cretaceous–Tertiary (K-T) boundary, 65 million years ago, was the last major event, although four smaller tektite-forming collisions have happened since. The last recorded large impact formed the Australasian tektite-strewn field. It occurred somewhere in Indo-China or in the South China Sea 770,000 years ago.

The Cambrian explosion [8]

Many reasons have been advanced to account for the apparently abrupt appearance of hard-shelled and skeletal fossils. We see the results as trilobites and other species preserved as fossils in strata of Cambrian age, as well as the marvellously preserved soft-bodied forms of many extinct animals in the Mid-Cambrian Burgess Shale, which was then mud on the ocean floor, but would eventually become part of the Rocky Mountains of British Columbia.

Among the prime suspects for the "explosion" are the recovery from the snowball Earth and the rapid rise of oxygen. Perhaps it was simple competition among the many soft-bodied forms that drove the development of bilateral symmetry, bones, hard shells, eyes and the ability to move quickly. However, this development seems to have been spread over perhaps tens of millions of years. Before the Cambrian, there is abundant evidence in the Late Proterozoic (Ediacaran) of multicellular life on the sea floor, living off the mats of bacteria. Perhaps some developed the ability to move, evolved sight and ate the others.

Whatever the cause, the Cambrian Explosion marks another improbable step in the evolution of life. There are several. These include the origin of life itself, the development of a genetic code, the proliferation of the simple bacteria and archea that thrived during the Archean and the development of the multi-celled eucaryotes early in the Proterozoic. None of these events seem related, although they are roughly coincident with geological and atmospheric changes. But false correlations are common in the natural sciences.

CONTINENTS

The continental crust of the Earth that most of us live on is of unique importance because it formed the platform above sea level on which the later stages of evolution occurred, so it is important to enquire how it came to pass [9].

Familiarity with our own crust has perhaps obscured how remarkable it is. Although the continental crust, about 40 kilometers thick, is less than 0.5% of the mass of the Earth, it contains a surprising amount, about one third, of the total Earth budget of many elements present in trace quantities, such as the heat-producing elements. There is no evidence on the Earth for that enduring myth of geology, a primitive world-encircling crust of granite. Our common granite, which decorates so many city buildings, turns out to be difficult for rocky planets to make, so the production of our granitic crust is unique in the solar system. It is the end product of three or more stages of distillation from the primitive rocky mantle, a consequence of plate tectonics. The continents have grown slowly and in an episodic manner throughout geological time and are now at their greatest extent. This process of forming the terrestrial continents is clearly inefficient. The Earth has transformed less than 0.5% of its volume to continental crust of intermediate composition and less than one fifth of 1% of its volume into the upper granitic continental crust in over 4000 million years. No company in the business of manufacturing continents would stay in business with that sort of record.

Rarely, geological processes conspire to concentrate rare elements, normally present at a few parts per million, or billion, into ore deposits. So the crust is rich in mineral deposits that contain these scarce elements that we find so useful in building a technical civilization. An often overlooked factor in the development of a technical civilization on a habitable planet is the need for a supply of the rarer elements, such as Ag, Au, Cu, Mo, Pb, REE, Sn, W and Zn. Ore deposits of these and many other technically useful elements are provided through the operation of plate tectonics. Color television screens contain the rare earth element, europium, which fluoresces to provide the red part of the image. Europium is present in the Earth at a level of only one tenth of a part per million, and rarely (hence the term Rare Earth) occurs in minable amounts. How many other planets might reproduce the endless geological cycles that have concentrated this element? Here, we can extract enough of it from exotic mineral deposits (even so costing several thousand dollars per kilogram) to spread it around the world in television sets.

Hydrothermal activity associated with volcanic activity, a primary mechanism for concentrating elements, is widespread.

However, only in a small number of sites are the local geological conditions appropriate to precipitate these metal-rich fluids to form minable ore deposits, as any economic geologist will tell you. Because of these accidents, the continental crust over geological time has become rich in mineral deposits that contain concentrations of rare elements that we find so useful in our complex technology. Thus the presence of the continental crust, developed through plate tectonics that are a unique feature in our planetary system, is a critical factor in forming this habitable planet.

So a habitable planet does not only need a favorable temperature, water and an atmosphere of oxygen. It needs deposits of copper, rare earths and lots of other things that we take for granted, but which come to us courtesy of plate tectonics. In contrast, the crust of Venus looks like a prospector's nightmare. The conditions for the production of extensive granitic crusts are unique to the Earth among our rocky planets. No crusts similar to the continental crust of the Earth have formed on the other inner planets, where the surface rocks are typically basaltic [9]. The significant feature about the Earth, in contrast to the other inner planets, appears to be the presence of liquid water at the surface, coupled with plate tectonics, which enables recycling of the oceanic crust through the mantle. It is this process that permits the slow production of the continental crust. In other planets the absence of plate tectonics leads to the persistence of barren basaltic plains, such as we observe on other planetary bodies and the Moon.

GLACIATIONS [7]

The reason for discussing these here is that their causes seem to be connected with several processes important to making Earth a habitable planet. Important factors include the relative amounts of methane, carbon dioxide and oxygen in the atmosphere, the movement of continental plates that causes change in oceanic circulation, the uplift of great plateaus such as Tibet, volcanic activity and wobbles in the orbit of the Earth. The latter are famous as Milankovitch

Cycles. They are responsible for the waxing and waning of cold periods within an ice age, but are not a primary cause.

Five great glacial periods have engulfed the Earth. The first recorded, discussed above, was the Huronian, which occurred shortly following the period of major geological change and the development of continents that marked the beginning of Proterozoic time. This was followed towards the close of the Proterozoic by the aptly named Cryogenian, which occurred between about 800 and 600 million years ago. It was another time marked by a "Snowball Earth" and a dramatic rise in the oxygen content of the atmosphere to present-day levels and the greatest variations ever recorded in the ratios of the heavy carbon isotope (with mass 13) to the lighter carbon isotope with mass of 12. This, the "Shuram Event", is generally thought to have been caused by the extinction of most photosynthetic life during the deep freeze. Shortly afterwards, precisely marked at 542 million years ago, the first complex fossils appeared. This event marks the base of the Cambrian Period and the beginning of the excellent fossil record that continues to the present.

The concept of a Snowball Earth has become well accepted. There is much evidence for glaciation at latitudes near the ancient equator. Once the Earth was frozen over, the white surface reflected sunlight so that a permanent frozen state might result. How then did the surface of the planet thaw? The answer in the model is that volcanoes continued erupting so building up carbon dioxide in the atmosphere. The resulting high temperatures in the greenhouse atmosphere melted the ice sheets. The excess carbon dioxide then produced acid rain which dramatically increased the weathering rate. Calcium and magnesium released from the rocks combined with the CO_2, resulting in the formation of massive deposits of limestones. These, the so-called "cap carbonates", are found immediately overlying the glacial deposits. This rock record showed that there had been a rapid change from icy to tropical climates, a previous geological conundrum resolved by the model.

Extensive ice ages occurred subsequently near the end of the Ordovician (the "Saharan") and later, well exposed in southern Africa, is the "Karoo" glaciation. This occurred at around 300 million years ago during the Carboniferous and the Permian. Ice sheets covered the ancient southern continent of Gondwana, now later split into the separate plates of Africa, South America, India and Australia through the operation of plate tectonics. The earlier development of land plants was perhaps responsible for that ice age. The plants produced an increase in oxygen production and a decrease in CO_2 levels, resulting in the cooling of the Earth.

During the intervening periods, the Earth was probably ice-free, even at the poles. About 20 million years ago, the Earth began to cool again as Australia drifted north, opening the oceanic circulation around Antarctica that now drives most of the weather in the southern hemisphere. The end product was the period of ice ages spanning the last 2 million years. Now we are enjoying a mild interglacial period that has lasted several thousand years, during which the current civilization grew. This is the 6000 years that Archbishop Ussher calculated.

The crust under the oceans

Rocky planets in our system produce crusts of basaltic lava and our planet is no exception. Enormous volumes of lava are generated over geological time. On our planet, the crust is erupted at the mid-ocean ridges and spreads outwards, a process called sea floor spreading. The discovery of this movement of the sea floor was one of the triumphs of geophysics in the 1960s. It was the smoking gun that showed that the crust on the Earth was mobile, leading to the notion of plate tectonics. Now it is well established that the sea floor moves at rates varying between 10 mm and 22 cm per year.

MORE ON PLATE TECTONICS

The process of plate tectonics has the useful properties of building continents and of forming ore deposits useful for advanced civilizations, so enabling this discussion to take place. But this process is unique to the Earth among our planets and maybe elsewhere.

The driving mechanism for the plate movement is mostly the pull of the subsiding oceanic crust, known as slab pull. How do these 100 km-thick plates move? They slide on a weak zone about 100 km deep in our mantle. What is it and why is it there? It even has a descriptive name, the asthenosphere. This zone is due to the presence of water that weakens the stiff mantle rocks. The water is there because the particular minerals at that specific depth and pressure cannot retain it within their crystal lattices. On such fine detail does the operation of plate tectonics depend.

There have been long disputes over when plate tectonics began, with some geologists arguing for its appearance in the Hadean, over 4000 million years ago. But the consensus is that modern style plate tectonics developed over several hundred million years in the Late Archean. The first definite evidence appears around 3000 million years ago and the process seems to have become firmly established early in the Proterozoic. At the transition between the Archean and Proterozoic, both the volume of the crust and differentiation within the crust underwent significant changes. Sir William Logan initially recognized this fundamental distinction in Canada between the Archean and the Proterozoic, in 1845, when he drew attention to the "great unconformity" between the Huronian sedimentary sequence and the underlying "granitic" basement in Ontario.

This unconformity marks the profound boundary between the Archean and Proterozoic Eons, and reflects fundamental changes in the way in which the continental crust evolved. However, the processes that were responsible extended over several hundred million years in different regions, to the despair of workers attempting to delineate precise geological boundaries. Thus, although the formal boundary is set at 2500 million years ago, nature rarely makes such sudden changes, except for catastrophic events such as the asteroid strike that occurred at the Cretaceous–Tertiary Boundary. So the switch from Archean-style crustal development to our familiar plate tectonic regime extended from around 3000 million years ago to well into the Proterozoic.

WATER ON THE EARTH [10]

Everyone is impressed by the great abundance of water on the surface of the Earth. However, the amount of water is very small in cosmic terms. The solar nebula in the neighborhood of the Earth became as dry as the Sahara, as water and other volatiles were swept out by the early active Sun to 3 or 5 AU (Figure 8). If the Earth had its proper share of what was present in the original disk of dust and gas, it would have several thousand times more water and so have become that body so often imagined in exoplanet literature, a "water planet". This amount would have trebled the volume of the Earth and drowned us in a deluge that would have astounded Noah.

In contrast to the other inner planets, the significant feature about the Earth is the presence of liquid water at the surface. Titan among the satellites is the only other body with liquid on its surface, but this is methane. Water is the critical factor both in facilitating the process of plate tectonics and in allowing the oceanic crust to be recycled through the mantle. The water trapped in the crust beneath the ocean is eventually taken down deep into the mantle. There it plays a critical role. As the crust sinks down into hotter regions, the water is driven off, taking with it many of the more volatile elements. As this fluid rises, it triggers melting within the mantle. This lava, now full of volatile elements, erupts on the surface in spectacular fashion. The eruption of Mt. St. Helens was a typical example. These events are seen most dramatically in the great chains of explosive volcanoes that occur around the rim of the Pacific Ocean, at the locations where the oceanic crust slides back down into the mantle. Such processes supply the materials that go to make up the continents, along with our useful ore deposits. In our other planets, the basaltic lavas stay on the surface.

So if water is so scarce in the inner solar system, where did the streams, lakes, rivers and oceans that we admire come from [10]?

Surprisingly, this could be treated as a trivial problem, since the Earth currently contains only about 500 ppm water. This is such a small amount that the source of the water on the Earth could be ignored, except that we are here as a consequence, "the heir to all the ages" [11].

What about forming the Earth from wet planetesimals? Although these are favorite building blocks among some geologists, they are easily enough dismissed. Three observations support this contention. The dry composition of most common meteorites is consistent with the "snow line" model and suggests that the planetesimals forming the Earth sunwards of the asteroid belt were likewise dry. Secondly, many elements such as potassium and lead are much less volatile than water but are depleted throughout the inner solar system (Figure 4). It would be surprising if water somehow escaped the general depletion of ices and volatiles that affected the nebula out to several AU.

Then if the rocky planets were assembled from wet precursors, the initial water contents of the other terrestrials could be expected to be relatively uniform. But Venus and Mars are much less differentiated and much drier than the Earth. If they had been assembled from wet planetesimals, these planets might be expected to have retained a wet mantle that would have facilitated the development of plate tectonics, contrary to observation.

Where then did the water in the Earth come from? Deuterium (D), the heavy isotope of hydrogen, is twice the mass of hydrogen (H) and the D/H ratio varies depending on the source. The D/H ratio of our water matches that of the famous CI meteorites from far out in the asteroid belt and at least two Kuiper Belt comets have the same D/H ratio. Comets such as Halley from the Oort Cloud, however, have too high a D/H ratio to be viable sources.

Current models thus suggest that most of the water on Earth came from a few icy planetesimals late in the formation of the Earth. These originated in the outer asteroid belt, so our water was most likely derived in a "hit or miss" fashion from drift-back of icy bodies from the Jupiter region.

The significance for the Earth is that just enough water (500 ppm) was added to form surface oceans. Calcium released by weathering of silicate rocks was able to combine with the atmospheric CO_2 and precipitate as carbonates, or be taken up by organisms forming calcite or aragonite shells. The accidental incorporation of just enough water also allowed for the operation of plate tectonics and reduced the melting temperature of basalt that was recycled into the mantle, enabling the formation of continental crust. Without this accident, Earth might have resembled Venus.

It is sobering to contemplate the differences between those apparent twins, Earth and Venus, formed by the random sweep-up of planetesimals, 100 million years after T_{zero}. The distinctions seen in the geology of the surface, the absence of plate tectonics and the contrasting rates of volcanic activity are probably ultimately due to the differences in the amount of water in the planets. All these features are the result of chance events during the formation and evolution of the planet. They do not encourage the speculation that Earth-like, in contrast to Earth-mass planets are common. None of the planets in the solar system, nor the many satellites, resemble one another; all are different and could just as well be members of another planetary system. The message is that chance events have played a crucial role in the origin and evolution of the solar system. Venus remains as a cautionary tale for seekers after "Earth-like" planets, with habitability depending on apparently trivial differences in mass and water content.

THE TURBULENT HISTORY OF THE ATMOSPHERE

> "The evolution of Earth's atmosphere is linked tightly to the evolution of its biota" [12].

The atmosphere of the Earth has been through a series of events of staggering complexity. There is little trace of any primitive gases. If the Earth formed within the gas-rich disk, it would have captured a thick primitive atmosphere. The rare gases, like neon, would be a hundred times greater than the present atmospheric content. Accordingly, it appears that the gas had gone by the time the Earth got around to forming. As a final insult, large collisions while the Earth was being put together would have stripped away any primitive atmosphere. Thus the present atmosphere and oceans of the Earth appear to be entirely secondary in origin and so provide little information relevant to the formation of the Earth, even if they are crucial in relation to the development of a habitable planet. Degassing of a wet interior is often appealed to, but the planet was assembled from dry planetesimals and the water and other volatiles came late as a result of "hit or miss" encounters.

But the most significant event was the rise of an oxygen-rich atmosphere early in the Proterozoic. The oxygen destroyed the methane. Was this arrival of oxygen in sufficient abundance not only to oxidize most of the iron in the oceans, but also to cause the appearance of the complex cells, the eucaryotes? The oxygen content rose slowly during the Proterozoic. Then it rose rapidly to close to its present levels late in the Proterozoic, more or less coincident with the glaciations that led to the Snowball Earth.

The faint early Sun

The Earth has maintained a rather even climate for 4 billion years. The Sun at that ancient epoch produced about one-quarter to one-third less light than it does now. This is the famous "faint early Sun" problem. The astronomical theory is robust enough, firmly based on the physics of nuclear fusion of hydrogen to helium. One might therefore expect that the early Earth would have been a frozen wasteland that warmed up slowly through the ages as the Sun increased its output.

In contrast, the geological evidence is quite definite that running water, eroding the surface and producing water-laid sediments, has been present throughout these vast epochs. Various explanations have been offered to explain how the Earth managed to maintain an even climate despite the faint sunshine that the astronomers insist upon. Usually some kind of greenhouse effect is invoked. One possibility for keeping the Earth warm throughout the Archean was that bacteria were producing large amounts of methane. But some CO_2 was present and a mix of both gases was probably the solution to the faint early Sun problem. However, the loss of methane may have precipitated the first recorded massive glaciation at around 2100–2400 million years ago.

Clearly a delicate balance has enabled the survival of the benign conditions that allowed life to continue.

LIFE ON EARTH

> "Even on this world, all of the available environments favor bacteria over scientists or theologians" [13].

Several steps appear particularly significant and equally improbable. The first was the origin of life itself and the subsequent proliferation of bacteria and archea. The second was the development of photosynthesis leading to the presence of oxygen in the atmosphere. This was closely followed by the emergence of the eucaryotes with complex cells. This development led to the appearance of complex organisms containing many cells. Next the explosion of life, with backbones, eyes, teeth and shells occurred at the beginning of the Cambrian Period. In spite of the continuing minor and major extinctions that sometimes came close to ending the experiment, life survived.

Although we may develop notions about our place in the universe, most are contingent. Thus the appearance of life and biological evolution on this agreeable planet does not predict the appearance of ants, elephants or *Homo sapiens*.

Is there a direction in evolution? The Burgess Shale, originally a muddy sea floor in Middle Cambrian time, 530 million years ago, has preserved a fauna of soft-bodied animals [14]. Such preservation is exceedingly rare in the geological record, for obvious reasons. Mostly, only the hard parts survive the many perils, such as being eaten, before becoming preserved as fossils. The preservation of the entire community now found in the Burgess Shale seems to have been due to a geological accident. A submarine landslide swept over the sea floor burying all these strange creatures. Only a few examples of such complete preservation are known in the entire geological record.

Apart from containing many well-known fossils, such as trilobites, what is interesting about the Burgess Shale is that it contains animals belonging to no known phylum. Phyla are the major divisions of the Animal Kingdom and there are currently somewhere between 20 and 32 of them, depending upon which biologist you talk to. The Burgess Shale contains another dozen or so organisms that are so distinct that they deserve to be classified as distinct phyla, depending upon which paleontologist you talk to.

These bizarre (to us) forms never appear again in the geological record. They have no modern counterpart. The grotesque animals that we can inspect, for example in the extensive collections in the National Museum of Natural History in Washington, DC, represent failed evolutionary experiments.

Thus the present millions of species represent only one set of possible life forms. Evolution could have wandered off in another direction with results impossible to predict. Even so, the diversity of those that we now observe is bewildering. The past forms were more so. We regard elephants with their long and useful trunks as remarkable and perhaps unlikely beasts. However, our two types of elephant (Indian and African) are the only survivors of 300 species of the formerly extensive family of the *Proboscidea* that included the giant woolly mammoths from Siberia and the various dwarf elephants of Crete, Cyprus and Malta.

Among the nine or ten crucial changes in evolution that might be mentioned, the development of flowering plants rates highly. These include the grasses that embrace food plants such as wheat, making them critical to our development [15]. This advance occurred very late, in the Mid-Cretaceous about 100 million years ago, although plants had colonized the land over 300 million years earlier in the Middle Ordovician or Silurian Periods.

But even on this most suitable of planets, there was nothing pre-ordained about the emergence of *Homo sapiens* on the plains of Africa. Three separate continents were available on the Earth on which the later stages of the evolution of land animals could develop. All of these vast areas shared the benign conditions on this planet that make it such a comfortable environment for life [16].

When animals first invaded the land in the Early Paleozoic, the continents were united into a single landmass, Pangaea. During the next few hundred million years, as plants and animals evolved and the dinosaurs became dominant, this great crustal mass began to split up, due to the rising of plumes of hot rock from deep within the mantle. These plumes seem to occur randomly, at least in our current understanding. The large southern continent, Gondwana, split away from the northern landmass of Laurasia. Gondwana in turn slowly fragmented into Australia, Antarctica and South America, leaving Africa in isolation.

So were formed the three great continental masses on which the later evolution of land animals proceeded independently. Australia, isolated from the rest of the world, produced weird marsupial animals. The South American monkeys, primates like us, never left the trees. Africa which in the meantime had rammed into Europe, alone managed to produce our unique species.

But that event was a consequence of geological and climatic accidents. The crust of East Africa, originally flat, was elevated due to hot plumes rising in the mantle beneath the crust in a region of tropical rainforest inhabited by primates. One plume rose under Kenya and another under Ethiopia. Between them, they split the rigid continental plate as it was uplifted, so forming the Great Rift Valley about 5 to 10 million years ago that now divides East Africa. In due course, East Africa will separate and drift away to the southeast as a separate continental plate, forming a new ocean in its wake.

Previously, the continent had been low-lying and covered with dense tropical jungle, where our primate ancestors lived amongst the trees. But the formation of the Rift changed the landscape into one of deep valleys, mountains and plateaus. The mountains that rose up to 4 kilometers, caused rain shadows, changed wind directions and turned the dense jungle, in which food was easy to find, into open savanna woodlands. Although the gorillas and chimpanzees still live in the jungles to the west, our ancestors were bold enough to venture into the open woodlands. Here it was useful to become two-legged, upright enough to see over the grass with binocular vision and agile enough to climb trees to escape predators (no one could outrun the large cats).

Climate changes followed as the Earth sank into the Ice Age. Several cold glacial periods were interspersed with warmer mild interglacial periods, such as we now experience. Superimposed on an overall drying trend, extremely wet and very arid cycles alternated. The great East African lakes that filled the rift valleys dried up periodically. All these dramatic shifts provided a stimulating environment for *Homo* and changes in species (e.g. emergence of *Homo erectus*) seem to have occurred during such climate extremes.

Brain size increased dramatically during the past 100,000 years or perhaps earlier, but seems to have stalled for *Homo sapiens* about 20,000 years ago. The cave paintings by Cro-Magnon Man seem to represent a pinnacle of artistic achievement. Brain capacity does not seem to have increased and has perhaps decreased since then. It is limited by the female body structure. Legs need to be close together for ease of walking. But this results in a small pelvic opening and a restriction to the size of the birth canal. But intelligence needs a large brain, making birth difficult. This explains the popularity both of midwifery and of Caesarean procedures, a response to the injunction in Genesis that "in sorrow thou shalt bring forth children". Although the first such modern operation took place in 1881, it now accounts for over a quarter of all births.

Until some alternative skeletal structure evolves that will allow a wider birth canal without restricting bipedalism, *Homo sapiens* has reached a dead end in evolution [17]. Indeed, even our physiology is much better adapted to a hunter-gatherer existence, with individuals living for 40 or 50 years. No need then for spectacles, dentures, joint replacements, hearing aids or the other benefits provided by science.

But without this random geological process, changing the landscape of East Africa due to unseen forces deep within the Earth, evolution among the 30 species of the genus *Homo* might have taken another turning. We might still be foraging in the jungle and not discussing the problem. So when everything else in the environment was perfect, random geological and climatic processes determined the development of intelligent life on this planet. The development of such life here seems governed by the same lottery-like chances of the formation and evolution of life, or of the planet itself in the first place.

But the most well-endowed planet is not immune to accidents along the way in the form of asteroidal and cometary impacts. Without the gravitational shield of Jupiter, this planet would be uninhabitable, although everything else is in balance. Even so, a large enough impact may devastate a planet. Such events, if they do not extinguish life completely, might rewind the evolutionary clock back to a primitive level where it may stay with the bacteria, or perhaps branch off in a different direction, as Stephen Jay Gould has argued. Evolution, lacking predictive power, has no necessary direction or plan to produce us as its ultimate achievement.

Such views are contrary to the notion, espoused most widely by Simon Conway Morris [18], that our emergence on this perfect planet was the result of convergent evolution. But primate evolution produced the genus *Homo* only in one location. Continents had split and drifted before, carrying, for example, the ancestors of the lemurs to Madagascar. But without that unique coincidence of plate tectonics and biology in East Africa, our species would never have arisen and the Earth would have continued along its heedless way.

Extinctions

> "No fact in the long history of the world is so startling as the wide and repeated extermination of its inhabitants"[19].

Of the species that have ever lived on the Earth, at least 99.9% are extinct. Species mostly have a relatively short life on geological timescales. Estimates typically vary from 100,000 years to 4 million, leading to estimates that about 50 billion species have existed on the Earth, although only one acquired high intelligence. Few gamblers would be interested in odds of one in 50 billion. But it is an interesting number as it is perhaps the number of planets in our galaxy. Our current population of between 10 and 30 million species displays a marvellous diversity in adapting to our myriad of environments. Some species of course live for much longer and have survived effectively unchanged for immense periods of time. The modern brachiopod genus *Lingula* is a classic example. It is similar in appearance and structure to its ancestor, *Lingulella*, that lived over 500 million years ago during the Cambrian Period. Sharks, which currently number about 300 species, first arose during the Silurian Period. Having successfully occupied their ecological niche, they retained their basic shape through a succession of different species for the next 420 million years.

There are many minor extinctions throughout the geological record. Evolution takes its toll, discarding species so rapidly that they have long formed useful markers for geological strata. Thus the rapid evolution and extinction of the tiny foraminifera allows geologists to divide up the rock succession in the Tertiary into zones of a few million years each.

Occasionally a combination of circumstances brings a dramatic increase in the rate, so that most species go extinct in a geological instant. Five such "major extinctions" are known in the fossil record in the past 500 million years. The first recorded occurred in the Ordovician, about 450 million years ago, followed by another in the Devonian, about 370 million years ago. Another major loss of species marks the end of the Triassic Period, which left the dinosaurs to dominate the land.

The disaster at the end of the Permian Period, 252 million years ago, dwarfs all the other catastrophes including the extinction at the Cretaceous–Tertiary boundary. Sometimes referred to as the "Great Dying", it came close to extinguishing life on Earth. Seventy percent of vertebrate families on land and 90% of species in the oceans became extinct. Wildfires were widespread. All this is less clear in the geological record than the episode that removed the dinosaurs. Strata at the Permian–Triassic boundary are less extensive than those that mark the end of the Cretaceous, for which we can study complete sections recovered from drill cores in the deep oceans. The sea floor of Permian times has vanished back into the mantle, carrying with it much of the evidence. But the discovery of sites in China has allowed precise dating of the event so that the extinction seems to have occurred abruptly, within 200,000 years. No evidence has been found of an asteroid impact. The great volcanic eruptions that produced vast expanses of basalt in Siberia that are close in time are often blamed. But a massive release of carbon dioxide and methane of organic origin into the atmosphere is favored in current models.

The extinction of many species at the boundary between the Cretaceous and Tertiary strata, 65 million years ago (the K-T event), has often been remarked upon. The successful dinosaurs, who had held sway for 160 million years, vanished along with a great multitude of species. What is particularly striking is that the extinction occurred at a geological instant. Where the geological strata are well preserved, as in drill cores from oceanic sites, the wipeout of species occurs at a knife-edge. No slow geological change in the continents, atmosphere or oceans can produce such a sudden catastrophe. This disaster left open many evolutionary niches to be exploited by the survivors. The mammals emerged and took over as the dominant land-dwelling species, leading to the eventual development of *Homo sapiens*.

The scientific consensus is that the ultimate cause was the impact of an asteroid between 10 and 15 kilometers in diameter. Something the size of Mt. Everest, travelling at least 20 kilometers a second, hit the Earth. The evidence for this impact at the end of the Cretaceous Period is particularly strong. The most striking and first discovered fingerprint was a worldwide spike of the rare element iridium in the thin layer of clay that marks the end of the Cretaceous in the geological record. Iridium is rare in the Earth's crust but common in meteorites. Next to be found in the clay, again over most of the world, were grains of quartz and feldspar that showed fractures due to a massive shock as though they had undergone a giant hammer blow. A lot of melted material that we now find as glassy fragments was sprayed around. Another team found soot from wildfires ignited by the event. There was so much soot that most of the forests on Earth must have burnt to supply it. All of these observations strongly support an impact event. No geological process, such as a great outpouring of lava, is capable of accounting for these facts.

The center of the impact is near Chicxulub on the Yucatan Peninsula in Mexico. The crater, or rather a ringed basin, formed by the explosion was at least 200 kilometers in diameter. Over the next 50 million years, it was slowly buried under limestone that was deposited in the warm Caribbean Sea. The asteroid struck a particularly deadly location. It dug in to several kilometers of strata containing carbonates and sulfates. These had formed by evaporation, in a bay of the ocean that had intermittently dried up and been flooded with seawater. The result was that hundreds of billions of tons of sulfur dioxide and carbon dioxide were blasted into the atmosphere. The sulfur dioxide combined with water vapor to form sulfuric acid. The resulting acid rain killed all near surface marine life that had shells of carbonate, easily dissolved by the acid. These included the tiny foraminifera that had lived in their countless billions. Nearly all of the species vanished in a geological instant.

Even those species, such as diatoms, with shells of more resistant silica, did not survive, probably on account of the "nuclear winter" effect. Several months of darkness were caused by the combination of dust and smoke in the atmosphere. First it was too cold and then it became too hot. Following the cold and dark interval, temperatures rose due to the large amount of carbon dioxide injected into the atmosphere from the carbonates in the target area, contributing further environmental stresses. These probably disrupted the entire food chain and so were also responsible for the extinction of the giant marine reptiles, such as the plesiosaurs whose graceful forms we admire in museums.

The present rate of extinctions is very high, due to human activities. Does this qualify for the sixth great extinction that the Earth has seen during the past 540 million years? Certainly the current rate of extinction is higher than that suggested by the record of the fossils. If the loss continues as at present, we will achieve the sixth mass extinction within a few hundred years, or effectively instantly in geological time. This has led to proposals that *Homo sapiens* is changing the geological scene so fundamentally that a new period has begun, the Anthropocene, currently thought to have begun around 1850 CE. Is this a new expression of the ego of our species? Certainly if our species vanished, it could take many millions of years to erase our traces [20].

THE MOON

Pre-Apollo visions

> "A surface cannot be characterized by its portrait. The Moon remained inscrutable at all scales" [21].

The study of the Moon produced so many insights into the history of our planetary system that it is useful here to provide an extended discussion about our satellite. The Moon is such a familiar object that it is often accepted just as being there, part of the "scheme of things". Laplace commented that it was an ancient view that "the Moon was given to the Earth to afford it light in the absence of the Sun" [22], although due to some oversight, this convenient arrangement works for only half the time. The Moon is in plain sight, accessible to naked-eye observation and has spawned a wide variety of cultural and religious ideologies, as well as serving as primitive calendars. The Ancient Greeks had established that the Sun was much further way than the Moon, despite their identical size in the sky.

The dark areas that make the features of "The Man in the Moon" are obvious to everyone, while a low power telescope or even binoculars reveal the craters that so impressed Galileo 4 centuries ago. He saw that the Moon was mountainous, thus fulfilling the prediction of Anaxagoras that the Moon was stony. History tells us that truth is often dangerous and the philosopher was banished from Athens for this heresy. Galileo had similar problems.

William Herschel, the discoverer of Uranus, thought that our satellite was inhabited and reported having seen cities. Even as late as the 1960s, some workers continued to postulate the existence of microbial life beneath the surface.

Once Newtonian mechanics was understood, the lunar orbit, mass and density were established, informing us of the much lower density of our satellite relative to the Earth. This major difference in density posed a serious dilemma for workers seeking to understand the origin of these two closely associated bodies. The low density of the Moon that was in startling contrast with the high density of Mercury, told us that our satellite was likely depleted in metallic iron relative to the Earth. Its curious orbit, inclined neither to the equator nor in the plane of the ecliptic, but at 5.1° to the latter, was another puzzle amongst many. Then the Moon did not have any discernible atmosphere.

It also displayed two distinct surfaces, unfamiliar to our terrestrial experience. The most extensive area was white and covered with immense craters of debatable origin, associated with some arcuate chains of mountains. These elevated regions were often supposed to be granitic, from analogy with the terrestrial continents. The other major surface feature that often occurred in semi-circular patches was smooth and dark, with many fewer and smaller craters. It was also solid, although early workers thought they discerned seas, hence the term maria for these features. These smooth surfaces were variously interpreted as dust, sedimentary rocks, dried-up lake beds or asphalt. More perceptive observers thought that these extensive plains were probably composed of lava, although the absence of terrestrial-like volcanoes was a puzzle if the flat plains were indeed volcanic in origin.

The Moon was large for a satellite. All the moons of comparable size that occur around the giant planets are relatively tiny compared to their planet. None of the other rocky planets had moons, discounting the two insignificant satellites of Mars. Why did Venus, our "twin planet", lack a satellite of any size? Why indeed was the Moon, so closely associated with the Earth, clearly distinct?

The fact that the full Moon is bright from limb to limb, rather than becoming duller toward the edges, as might be expected from a sphere that was reflecting sunlight, has often been commented upon. It was finally understood that the surface is broken up into tiny fragments by the impact of meteorites. Thus it is covered with a myriad of reflecting surfaces, like a series of bicycle reflectors in the sky. This may explain how the ancient Greeks arrived at their concept that the heavenly bodies were made of shining crystal.

Among other controversies, the notion that tektites had been derived from the Moon enjoyed widespread support in the 1960s. It was often thought that tektites were samples from the lunar highland crust on account of their supposed "granitic" composition. The concept that the lunar highlands were granitic was based on analogy with the terrestrial continents. But in around 1960 detailed examination of their composition revealed that tektites were derived not from granites, but from sedimentary rocks that were unlikely to be found on a body lacking an atmosphere.

Richard Dawkins (b. 1941), the biologist and modern defender of Darwin's "dangerous idea", has pronounced that "the Moon is simple" [23]. Indeed, it might appear so to a biologist attempting to account for the development of the eye, but the nature, composition, evolution and origin of the Moon baffled scientists. Before the lunar missions, the heroic efforts of cosmologists and a wide variety of experimental scientists all failed to provide an adequate explanation for either the composition, history or existence of the Moon.

Some physical and orbital properties became well established before the Apollo Missions, although data for the crucial moment of inertia was lacking. With the benefit of hindsight, it is clear, however, that several key facts, among them the low density of the Moon, the strange orbit, and the rapid spin of the Earth–Moon system, were already known, awaiting integration into a coherent theory. Debate raged over the nature of the highlands, the maria, the craters and the composition and the origin of the body itself. Thus the Moon before the Apollo landings was, although clearly visible, immensely puzzling to terrestrial observers. The lack of data indeed encouraged endless speculation among a species looking for order and regularity in the universe. Nothing had been settled after 300 years of speculation. But all this was to change with the arrival on Earth of a few handfuls of samples.

Rosetta Stones

The Moon has played a central role in the development of theories of the origin and evolution of the solar system. This is not without irony, since it has proven one of the most difficult objects to explain. It is in plain sight, accessible even to naked eye observation, as Harold Urey (1893–1981) who persuaded NASA to go to the Moon, was accustomed to remind us. It is an obvious first object to fit into theories for the origin of the universe, although one distinguished scientist remarked that it was so small that it could be ignored. The Moon was often thought to be a kind of Rosetta Stone [24], so that the general belief before Apollo was that we could discover much about the origin of the solar system by going to the Moon. This was a major scientific justification for the manned lunar missions, although political considerations were the real motivation.

The Moon eventually did provide us with a kind of Rosetta Stone for understanding of past history of the solar system, but not in the manner imagined by Earth-bound thinkers before the Apollo Missions. One of the principal conclusions of lunar studies was to demonstrate the importance of large impacts of asteroids, meteorites and comets in the early stages of the solar system. The evidence from the wide range of impact crater sizes led to the notion that a variety of objects of differing sizes existed, and that our rocky planets had formed from these rather than from fine dust.

The space missions provided crucial additional information on ages, chemistry, and the significance of cratering, in particular of the importance of large basin-forming impacts. The largest structures formed by such processes are basins surrounded by circular rings of mountains (Figure 10).

Such random processes could explain many features of the solar system, for example the various tilts of the planets, so that a new understanding of the origin of planets and satellites emerged. Once we properly understood the origin of the Moon with all its implications and connotations, then a clearer view of the rest of the solar system started to emerge and the somewhat untidy nature of the solar system began to receive a rational explanation.

The Apollo Missions

The Ranger, Orbiter and Surveyor missions, designed to provide information about the nature of the lunar surface, preceded the Apollo 11 landing in 1969. The most significant in this context were the Surveyor Landers. They were designed primarily to test the bearing strength of the surface, because of the notion that the maria, on which Apollo 11 was to land, might be full of loose dust. However, they found a firm surface and also provided some chemical data suggestive of basalt.

But it was the samples from the Apollo Missions that both resolved many problems and presented scientists with new dilemmas [25]. The surface of the Moon is covered with a blanket a few meters thick of rubble and dust from the impacts of meteorites, so that it presents a rounded rolling surface to the astronaut. The absence of familiar landmarks makes it extraordinarily difficult to judge distances. There is a surprising amount of relief, over 16 kilometers between the highest and lowest point. This is only a little less than the extremes of 20 kilometers on the Earth between the top of Mt. Everest in the Himalaya and the Challenger Deep in the Marianas Trench of the western Pacific Ocean. On the Moon, the rugged terrain is the result of giant impacts, gouging great basins, rather than the forces of plate tectonics that continue to shape the surface of the Earth.

The Moon has a thick crust, about 9% of planetary volume, that formed rapidly following the formation of the Moon at about 4460 million years ago, 100 million years after T_{zero}. The crust, about 50 km thick, varies widely, reaching 100 kilometers on the far side, on a body whose radius is only 1738 kilometers. The continental crust of the Earth, in contrast, about 40 km thick, is relatively much smaller and has grown slowly in fits and starts throughout geological time.

The lunar highland crust is different in composition from the interior and contains a large proportion of feldspar which is responsible for the white color of the lunar highlands. It is complex in detail, mainly because the rocks were smashed up by meteoritic bombardment. However, their chemical composition has survived to tell its tale. The generally accepted model is that the crust formed as crystals of feldspar formed like icebergs on the surface of a molten Moon.

The fact that the mountain ranges on the Moon lay along arcs of circles was a great puzzle to early investigators. Now we understand that they form like giant ripples of rock as huge bodies slammed into the Moon (Figures 10, 17). A major insight from the study of the Moon was this evidence for massive impact cratering early in the history of the solar system. One of the most striking features of the lunar surface is the evidence of meteorite impacts at all scales, from large basins, ranging from hundreds to over 1000 kilometers in diameter, with concentric rings of mountains, down to tiny micron-sized pits caused by micrometeorites hitting grains lying on the surface. Although the larger craters were long thought to result from volcanic activity, their origin due to the impact of asteroids, comets and meteorites, was established not long before the Apollo Missions, following much controversy over impact versus volcanic shaping of the face of the Moon.

Meteorite impacts have occurred at a much slower rate on the Moon since the termination of the Late Heavy Bombardment. The youngest such major event on the Moon was the formation, about 100 million years ago, of the crater Tycho, 85 kilometers in diameter. Tycho formed due to the impact of a small mountain-sized body, a few kilometers in diameter. Material ejected during this impact forms the bright rays, which extend across the visible face, and are such a spectacular feature of the full Moon, particularly when viewed through binoculars. This large impact on the Moon is not far removed in time from the similar catastrophe on the Earth that wiped out the dinosaurs and much else.

There is a great contrast between the long-term survival of craters on the Moon and their rapid destruction on the Earth. About 35 million years ago, an asteroid struck at what is now the mouth of Chesapeake Bay on the eastern seaboard of the United States. It dug a crater 90 kilometers in diameter, about the size of Tycho. The collision sprayed out some glassy fragments, which are now found as tektites as far away as Texas. In contrast with craters on the Moon, which remain untouched until destroyed by later impacts, the young crater in Chesapeake Bay was rapidly filled by sediment and covered from view. It has only recently been discovered by a combination of geophysical probing of the subsurface structure and by study of samples obtained by drilling. Meanwhile, Tycho, which is about three times older, sits untouched, with its beautiful rays radiating across the face of the Moon.

Basaltic lavas (the maria) form the familiar dark features of the "Man in the Moon". Lavas rise to the surface of the Moon and elsewhere because the liquids are less dense than the surrounding solid rock. Lavas on the Moon are more common on the near side. There they can more easily reach the surface because the crust is thinner. In contrast, they are rare on the far side of the Moon because they mostly fail to reach the surface through the much thicker crust. The maria, although prominent visually, form only a thin veneer on the thick highland crust and constitute less than about 1% of crustal volume.

The lavas fill various impact basins and craters to differing levels. Over 25 distinct varieties have been sampled, so that they come from many distinct regions in the mantle of the Moon, in contrast to the uniform composition of the common basalts erupted on Earth at the Mid-Ocean Ridges. These terrestrial lavas have much more uniform compositions than their lunar counterparts. The lunar lavas are also very ancient, mostly erupted between 3800 and 3200 million years ago, in contrast with the younger age of our MORB.

The lack of our familiar volcanoes on the lunar surface is a result of the very fluid nature of the iron-rich lunar lavas. This enables them to flow as easily as oil for hundreds of kilometers on slopes of only a degree or two, making remarkably flat plains. This was unexpected. Basalt lavas on the Earth are viscous, more like toffee than engine oil. The Moon contained many other surprises for geologists and geochemists. These taught us that every planet has a distinctive history.

The basaltic lavas were erupted from deep within the Moon. They formed by the melting of zones of differing composition. These zones formed during the solidification of the Moon, as various minerals precipitated as the Moon cooled. Some contained radioactive elements that slowly heated, eventually melting the more easily melted minerals, forming a liquid slush or magma. Because this melt was of lower density than the mantle, it rose to the surface and flowed out as lava. Some melts had entrained traces of volatile elements, that sprayed out as fire fountains, producing many tiny globules that astronauts found as the famous green and orange glasses.

There is good evidence for the existence of a small metallic core, but it forms only a few percent of the volume of the Moon. Iron and its associated elements (nickel, cobalt, platinum, iridium etc.) are between 20 and 50 times less than the traces on the Earth and are mostly in the core.

Among the other surprising results from the Apollo samples was the demonstration of ancient magnetic fields, now extinct. The most likely possibility is that the field was generated internally during the freezing of the tiny fluid iron core.

THE COMPOSITION OF THE MOON; IS IT A RELIC FROM THE EARLY SOLAR SYSTEM?

The first Apollo sample returned from the smooth basaltic plains of Mare Tranquillitatis found some unusual chemistry in contrast with our terrestrial rocks. The Moon is strongly depleted in the most volatile elements, such as lead and chlorine, as well as the moderately volatile elements such as potassium (Figure 4) and sodium and is enriched in the refractory elements, like calcium and aluminum, compared with the Earth.

This volatile depletion is well shown by the very low K/U ratio of the Moon, 2500, compared with 10,000 for the Earth or 60,000 for the original nebula (Figure 4, Figure 6). The Moon is more strongly depleted in volatile elements than the Earth, but not compared to the asteroid Vesta, and is completely dry except for trivial amounts of water (see below). The trace metallic elements, such as nickel and platinum, are depleted in the Moon to a similar extent as on Vesta. These elements concentrate in metallic cores, but this likely occurred in the core of the impacting body (Theia) that hit the Earth to form the Moon.

Vesta produced basalts with similar element patterns to lunar basalts, including strong depletions in volatile elements within a few million years of T_{zero}. Thus at least one asteroid lost volatile and siderophile (metallic) elements similarly to the experience of the Moon, although the metallic elements in Vesta are likely located in its core. Data from isotopes also tells us that the Moon lost volatile rubidium relative to refractory strontium very early on, much like the meteorites (eucrites) from the asteroid Vesta. This loss occurred at a very early stage in nebular history, close to T_{zero} (4567 million years ago).

Potassium has not undergone any isotopic fractionation (Figure 6), so that the potassium now in the Moon, like that in all other measured inner solar system material, including the Earth, has not undergone significant evaporation or condensation during formation of the Moon.

All this evidence strengthens the case that the impactor that formed the Moon had a similar history to Vesta and that the dry and volatile depleted composition of the Moon was not due to the collision that formed it. The loss of volatiles occurred at T_{zero} and was inherited by the Moon from the impactor. Thus the strange (to us) composition of the Moon was mostly established close to T_{zero} and is not a consequence of the Moon-forming impact. That little-understood event homogenized the oxygen isotopes and some other isotopes in the vapor cloud and is perhaps responsible for the enrichment of the Moon in refractory elements.

Water on the Moon

Water is essential to life as we know it, so there is an understandable interest in the presence of water elsewhere, even on a body as dry as the Moon. The NASA LCROSS Mission successfully detected water, following the impact of the upper stage of the launch rocket into one of the shadowed craters near the south lunar pole. These traces of water, trapped as ice in the lunar soils, were delivered from comets, or from the interaction of hydrogen from the solar wind hydrogen with surface minerals.

This evidence for traces of water on the surface needs to be distinguished from reports of water (present as the OH ion) in minerals derived from the interior of the Moon. The lunar rocks are dry and strongly reduced. No micas, amphiboles that contain OH, have ever been found. Ferric iron that would indicate oxidizing conditions is absent as well.

But trace levels of very volatile elements such as fluorine and chlorine have long been reported. As both elements are much more volatile than water, it should not have been surprising to any geochemist that a minute trace of water might be incorporated within the Moon along with trace (ppb) levels of many other trace volatiles.

During the crystallization of the lunar magma ocean, such elements were concentrated in residual fluids. In this process, any water would finish up in the residual melt (along with other incompatible elements). Minute traces of water were also trapped in mineral zones in the deep interior and are now found along with fluorine and chlorine in the OH sites in minerals such as apatite, a trace mineral in the basalts.

Water present in apatites from the Moon has a very variable isotope ratio (D/H), higher than that of terrestrial water, in contrast to the identical oxygen isotope ratios in both bodies. This tells us that the trace of water on the Moon did not come from the Earth and probably came from the impacts of a few comets while the Moon was still molten.

The Moon is drier than the Earth by factors between 10,000 and 100,000. The Moon is thus essentially devoid of water, unless one regards the presence of one or two parts per million as significant. The great publicity about traces of water, either on the lunar surface or in the interior of the Moon, seems to be due to the sacred cow status that has been given to water. While this is obvious from a human viewpoint, it obscures the fact that any water in the inner solar system arrived there by chance.

The origin of the Moon: early notions

The Moon is in plain sight, but defied efforts to explain it for hundreds of years. Attempts to explain complex natural phenomena call for a wide variety of skills and the Moon presented a particularly difficult example, one of the enduring questions. How did it form and why is it there? It is sobering to record that none of the models that had been erected before the Apollo Missions were able to account for it. Three centuries of theoretical reasoning were negated by a handful of samples.

Following the sample return, all pre-Apollo theories for the origin of the Moon failed for various reasons. Hypotheses in which the Earth captures an already formed Moon were abandoned. It turns out to be very difficult to capture the Moon into its present orbit around the Earth. In such a model, the curious chemistry of the Moon had to form somewhere far away. Putting problems out of sight does not solve them.

The similarity in density between the Moon and the Earth's silicate mantle fuelled speculation, dating back to George Darwin (1845–1912), a son of Charles, that the Moon had formed from the Earth's rocky mantle following core formation on the Earth.

Such "fission hypotheses" that derive the material for the Moon from the Earth's mantle, although popular, encounter two basic difficulties. The spin of the Earth–Moon system, although large, is insufficient by a factor of about four to allow fission to occur. A second objection is more telling. Fission models have become the most readily testable of all following lunar sample return, as they predict that the chemistry of the Moon should bear some recognizable signature of the rocky mantle of the Earth. However, the composition of the Moon proved to be distinct. Although the Moon has a low metal content, its rocky mantle has 50% more FeO than that of the Earth. It was also depleted in volatile elements and enriched in refractory elements, such as aluminum. However to add further to the conundrum, the ratios of the oxygen and chromium isotopes turn out to be identical in both bodies.

Double planet models that form the Moon and Earth in association possess the twin difficulties of failing to account for the high spin of the Earth–Moon system, and of readily accounting for the density difference.

Yet another model formed the Moon from a ring of rocky debris produced by break-up of incoming asteroids as they came within the Roche Limit. This process was supposed to result in a ring of broken-up rock debris around the Earth. Their tougher iron cores stuck together and crashed into the Earth. This sounds more like a process for making rings rather than satellites.

These last two models form the Moon as a natural process related to the formation of rocky planets and so should lead to the presence of moon-like satellites elsewhere. But they fail a crucial test. The absence of a comparable satellite around our twin, Venus, is the fatal flaw in these notions.

None of these theories accounted for the unique nature of the Earth–Moon system, for the strange lunar orbit and for the high spin of the Earth–Moon system. Like ships hitting an uncharted rock, they all sank from this defect. Uncommon objects like our Moon require a unique mode of origin.

THE GIANT IMPACT MODEL

Perhaps the most significant property of the Earth–Moon system is that its angular momentum is high compared to that of the other planets. This problem was pointed out by Al Cameron (1925–2005), who showed that this excess angular momentum had to have arisen through the collision of the Earth with an object a little larger than Mars. The high spin rate of the Earth and Moon cannot arise from a series of small impacts. However, one large impact could account for it. But, one had to leap several orders of magnitude beyond the scale of the great impact scar of Mare Orientale (Figure 10) to propose the single large impact hypothesis for lunar origin [26].

The required conditions in the model are for the impactor (Theia) to have a mass about 15% of that of the Earth (larger than Mars) and to hit at a glancing angle with a velocity of 5 kilometers a second. This body, as well as the Earth, is assumed to have formed a metallic core and rocky mantle before the collision. Both the Earth and the impactor are melted. The metallic core of the impactor separates from the rocky mantle and falls into the Earth within a few hours. The molten mantle of the impactor forms a disk around the Earth that includes some of the molten mantle from the Earth. Temperatures in the disk reach several thousand Kelvin. The behavior of materials under these conditions is not readily modeled.

The fact that the composition of the oxygen and other isotopes is identical in both the Earth and Moon is often used to support the notion that the Moon was derived from the Earth, rather than from Theia. This contrasts with widespread variations in isotopes in other bodies in the inner solar system. Clearly the oxygen isotopes were homogenized as a result of the impact, but the bulk chemistry was not affected and the composition of the Moon remains distinct from that of the Earth. But processes in such high temperature liquid–vapor systems might homogenize the isotopes without affecting the difference in bulk composition. We need much more investigation of the temperature conditions during the giant impact.

The impact event was also sufficiently energetic to vaporize much of the material that went to make up the Moon. This is often supposed to explain such unique geochemical features as the dry nature of the Moon and the extreme depletion of very volatile elements. However, there is no evidence that volatile elements were boiled off (Figure 6). The dry and volatile depleted nature of the Moon was inherited from the impactor and dates back to T_{zero}.

A final consequence of the giant impact model for lunar origin is that the event is energetic enough to melt the mantle of the Earth. Such melting appears to be an inevitable consequence of the accretion of large planets, but the effects on the geochemical evolution of the terrestrial mantle remain to be fully evaluated.

Of course, a prime requirement for this useful hypothesis is that there was a supply of bodies of the right size to hit the Earth. Fortunately for the model, there is plenty of evidence from the tilts of the planets for the previous existence of Mars-sized bodies in the inner solar system. Unique events are difficult to accommodate in most scientific disciplines. However, although the details of the

Moon-forming event are not predictable, massive collisions early on in the solar system were common. One just happened to have the appropriate mass and velocity and hit the Earth at the right angle to provide us with the Moon that has been such a source of inspiration to poets and princesses alike.

What was the date of this event? Although dates as early as 30 million years after T_{zero} were commonly suggested, these came from an erroneous interpretation of the hafnium-tungsten isotopic system. Current models suggest ages around 4460 million years ago, consistent with dates for the crystallization of the highland crust. Thus, formation of the Moon took around 100 million years after T_{zero}. This date represents the youngest possible date for the formation of the Earth as well.

EVOLUTION OF THE MOON

The broad features both of lunar composition and evolution are well understood and known better than for the Earth (Figure 17). The views about the Moon before the Apollo Missions cast their revealing light, had led to the view that the Moon was a primitive object. This idea arose because of its low density relative to the Earth. Thus it is ironic that much of its composition was established close to T_{zero}.

A consequence of the giant collision was that most, if not the entire Moon, was melted. This vast mass of molten rock has been termed the "magma ocean" and a highly energetic and rapid mode of origin, as provided by the Giant Collision model, is required to account for it.

The cooling and crystallization of this ocean of melted rock is understood in principle. Early-formed minerals such as olivine crystallized and sank. Feldspar, forming a little later, floated, forming a thick crust due to the low density of the feldspar crystals and the bone-dry nature of the silicate melt. "Rockbergs" of feldspar may have been swept together like icebergs by tidal effects or convection currents. This could account for the differences in crustal thickness between the near and far sides. Other plausible models suggest a

FIGURE 17 We now understand what we are looking at on the surface of the Moon. The contrast between the lunar highlands and the lunar maria is well shown in this view of Mare Ingenii on the lunar farside. The rugged white regions are the primary crust of the Moon, mostly made of calcium feldspar. Large circular craters have been punched into this crust by impacts. Millions of years later these holes have been flooded with basaltic lavas erupted from deep in the mantle, producing the smooth grey plains of the maria. These have a few small craters on their surface from later impacts. The large circular crater, filled with mare basalt, is Thomson, 112 kilometers in diameter, in the northeast sector of Mare Ingenii, which is 370 kilometers in diameter. The sequence of events that produced this scene from oldest to youngest, is (1) formation of white feldspar-rich highland crust, (2) excavation of Ingenii basin, (3) formation of Thomson crater, (4) flooding of Ingenii basin and Thomson crater with basaltic lava and (5) production of small impact craters on the smooth mare surface, including a chain of secondary craters across Thomson. (NASA AS15–87–11724).

pile-up of debris from the early collision that formed the gigantic South Pole Aitken Basin, which is 2500 kilometers in diameter and 13 kilometers deep. A tiny lunar core of iron formed in the center. This took in what little nickel, platinum, iridium and gold was left in the Moon after the core of Theia finished up in the Earth.

Samples from the highland crust, which is dominated by calcium feldspar, are mostly broken-up breccias as a result of the bombardment that continued as the crust formed. Fortunately these did not change the composition so that the chemical record can be read. Impact melts are common.

The lunar rocky mantle was fully crystallized within a few million years. Zones formed of different minerals, from which the lavas that darken the face of the Moon were derived, much later. The final dregs resulting from the crystallization of the Moon were highly enriched in those elements that could not fit into the common minerals. This material, referred to as KREEP from potassium (K), rare earth elements (REE) and phosphorus (P), was mixed into the crust by impacts and accounts for the very high concentrations of elements such as potassium and uranium in the crust of the Moon. However, in the absence of the recycling that accompanies plate tectonics, this material was not concentrated into ore deposits that we could mine.

The maria, although prominent visually, form only a thin veneer on the thick highland crust. As noted above, over 25 distinct varieties have been sampled. The Moon has an extremely thin atmosphere, consisting only of a sprinkling of atoms that are sputtered off the surface by radiation from the Sun.

THE EFFECT OF THE MOON ON THE EARTH

The Moon-forming and other massive collisions had several effects significant in making the Earth a suitable abode for life. Any early dense atmosphere was removed by the impact. The tilt of the planet, another consequence of the collision, produces the seasonal variations. The rather rapid rotation of the Earth, in contrast to the slow rotation of Venus, provides a day–night temperature and light variation less stressful to the development of life than much longer or shorter periods might be. The day has become a little longer over time as the Moon has receded from the Earth, so slowing the rotation of our planet by tidal forces as it retreated. About 1 billion years ago, when the Moon was about 35,000 kilometers closer, the day was about 18 hours long.

In contrast to that of Mars, which wobbles through as much as 60°, the Moon stabilizes the obliquity of the Earth because of its large mass and its closeness to the planet. According to Lascar and co-workers [27], the obliquity of the Earth would be chaotic with large variations in the absence of the Moon. Widely varying tilts would cause variations in the amount of sunlight that regions would receive. The associated changes in climate would presumably produce a stressful environment for life. So our satellite is a climate regulator for the Earth.

The atmosphere of the Earth is secondary, as any primitive atmosphere was removed by the lunar-forming impact. The Moon also causes tides in the Earth's oceans that provide a stimulating near-shore environment for life. All have implications for life and climate, so that the Earth may owe its habitability to the presence of the Moon.

All this adds more factors to consider in the search for habitable planets in other planetary systems. Do habitable planets need moons? In the context of this discussion, two questions without answers are, how crucial is the presence of the Moon in producing this habitable planet and how likely are such events in other planetary systems?

LIFE ON THE MOON

Christian Huygens (1629–1695) said 350 years ago that "the Moon has no air or atmosphere surrounding it as we have. I cannot imagine how any plants or animals whose whole nourishment comes from liquid bodies, can thrive in a dry, waterless, parched soil" [28]. Such opinions did not prevent Carl Sagan, 3 years after Apollo 11, from speculating that "early lunar conditions are not inconsistent with the production of large quantities of pre-biological organic matter", now buried in the regolith [29]. However, our technical civilization is capable of providing a comfortable environment there. Residing on the Moon, shielded by a few meters of the lunar soil, that is moved as readily as beach sand on Earth, would not be very different from living in Antarctica, except that oxygen and water would be needed. The polar regions seem the most likely site for an inhabited lunar base. Some sites at the poles on crater rims enjoy continuous sunlight. There one could have a permanent solar power supply, a consideration that will over-ride all other requirements in selecting a site. Other regions would have to endure 2 weeks of darkness each month.

7 Perspectives

> "If a person should ask my advice before undertaking a long voyage, my answer would depend on his possessing a taste for some branch of knowledge" [1].

At present there is a popular consensus in favor of intelligent life elsewhere in the universe. These notions are at least 2500 years old and rest on several assumptions, that have run as a common thread throughout history. Many lie outside the realm of scientific enquiry coming under the heading of "must be" arguments. The first is that the universe is infinite and so must contain planets identical to the Earth. This is often phrased as the "Big Numbers" argument. There are so many stars with planets that somewhere out there must be a replica of us. The second is that because life exists here, it must be common elsewhere. The third is that the development of intelligence is inevitable and happens elsewhere concurrently with, or more commonly in advance of, the evolution of life on Earth. These themes are addressed below under several headings.

THE PLURALITY OF WORLDS [2]

The discovery of many planets orbiting other stars, free-floating objects and the widespread occurrence of dusty circum-stellar disks, some with gaps in which planets are lurking, has raised once again in a dramatic fashion, the ancient question posed amongst others by Albertus Magnus in the thirteenth century: "since one of the most wondrous and noble questions in Nature is whether there is one world or many, a question that the human mind desires to understand, it seems desirable for us to enquire about it" [3]. One of the favorite current quotations of astrobiologists comes from Metrodorus of Chios (350 BCE). It is usually stated as "it is unnatural in a large field to have only one ear of corn and in the infinite universe, only one living world". These questions have been discussed under many headings for the past 25 centuries since Democritus, Epicurus and Metrodorus favored a multitude of worlds.

The concept of a multitude of habitable worlds has appeared historically under several headings such as "the plurality of worlds", "the principle of plenitude (abundance)" and most recently as "the principle of mediocrity", which states that our neighborhood is more or less typical of the rest of the universe [4]. These sentiments are currently common and seem to be a reaction to pre-Copernican views of the central importance of the Earth.

Among many others, Derham (1715) in his book *Astro-Theology*, appealed to the similarity among the planets in the solar system as evidence that they might all be inhabited. Similarity does not imply identity. That notion even has some difficulties with stars (most are red dwarfs), while the study of our solar system reveals that random events produced our system of planets. Although many people, including astronomers, are surprised by the diversity of exoplanets, our solar system informs us that no object in it is similar to another. Wide variations among exoplanets might therefore have been expected to be the norm.

Now that the question of whether there are planets around other stars has been resolved by sophisticated technology, we can examine how likely they are to resemble the Earth. Although we have only one habitable planet in our system, the discovery of these exoplanets has raised popular hopes of finding Earth-like planets elsewhere. It then takes only a little imagination to populate such agreeable sites with an interesting assortment of intelligent beings, with whom we might ultimately be in touch.

The question of the existence of other Earth-like planets is sometimes referred to as the term N_e, "the number of planets ecologically suitable for life" in the controversial Drake equation [5]. Here we have seven other planets, at least 160 satellites and an assortment of asteroids, centaurs, Kuiper Belt and Oort Cloud bodies that were produced along with the Earth from the primordial solar nebula. The question turns on whether this planet formed as an inevitable and natural consequence of planetary-building processes, or whether its formation and history were dominated by chance events. If the first scenario is true, then duplicate copies of our agreeable and habitable planet, although lacking in our system, might be expected to be common elsewhere. If the formation and evolution of our planet, as well as that of life and intelligence, has been dominated by random processes, then the presence of Earth-like planets elsewhere becomes much more of a lottery.

HABITABLE ZONES AND HABITABILITY

> "Of course, planets must be volcanically active and they must be endowed with adequate supplies of both water and carbon" [6].

Contrary to popular supposition, the universe is wildly inhospitable to life. It mostly consists of one hydrogen atom per cubic centimeter. Life as we know it is clinging to a thin film around a small rocky planet. Life clearly requires special sets of conditions, such as those provided by planets, both to get started and then to survive and evolve. The position in the galaxy also matters. Near the center, radiation levels seem too high for life. The outer parts of our galaxy are mostly poor in metals, stars there having less than 10% of the metal content of our Sun. Perhaps we are situated in a "habitation doughnut".

Not all galaxies are suitable environments in which Earth-like planets might form. Our small dwarf satellite galaxies, the Magellanic Clouds, are poor in metals and so are less likely candidates for planets. Heavy elements dispersed from dying stars have a patchy distribution so the universe is not well mixed.

One of the persistent fallacies in the debate about habitable planets is the confusion between necessary and sufficient conditions. Most discussions assume, for example, that we are dealing with

Earth-like planets with breathable atmospheres, the presence of liquid water on the surface and residence in a narrow habitable zone for several billion years. The latter is a restrictive constraint requiring a nearly circular orbit (i.e. surface temperatures between 0°C and 100 °C) (Figure 18).

Others take one parameter as sufficient. Thus, evidence that an atmosphere (e.g. Titan) or water (e.g. Europa) is present is sometimes considered strong evidence that life may be present. But many other unrelated factors are involved, as the French astronomer, Hervré Faye (1814–1902), perceived in 1885, "If it were possible to make a complete enumeration of these conditions [for life] which, in the majority of cases, are independent of each other, one would see that there are indeed few chances that they would be found united on any globe. Nature has consequently had to form a great number of worlds for that one habitable milieu to be produced... by a fortunate concourse of favorable circumstances" [7].

Other factors that may influence habitability may include the presence of a giant planet like Jupiter to act as a shield against meteorite, asteroid or cometary impacts. If Jupiter did not exist, or was smaller, the Earth would be bombarded with comets. The numbers of impacts on the Earth would be perhaps a thousand times greater than at present. Local areas, a few kilometers across, might experience a catastrophe several times a year, instead of once in a few thousand years. Collisions of the sort that killed off the dinosaurs might occur every million years rather than every several hundred million years. This bombardment might have had incalculable effects on the development of life, perhaps stopping it altogether. Even with the protection of Jupiter, life on this planet has come perilously close to being extinguished several times, although an impact has only been demonstrated once as the cause. It seems unlikely that our species, or life itself, could have survived such disasters without the protective shield of Jupiter. In the few million years that the genus *Homo* has been around, we would have experienced between 20 and 100 such impacts.

The 23.5° tilt to the plane of the ecliptic, another likely outcome of the lunar-forming collision or of a similar catastrophe, provides the seasons, a great stimulus to evolution, while the 24-hour rotation period, also favorable to the development of life, is a consequence of such chance collisional events.

Perhaps a large Moon is needed to stabilize planetary tilt. The presence of our Moon raises substantial tides in the Earth's oceans that also stimulate biological activity. The Moon is unique within the solar system and the particular parameters of the Moon-forming collision make its duplication elsewhere improbable. Although tilt-stabilizing moons might arise elsewhere, this assumes that the habitable planet is tilted at an agreeable angle, euphoria being a staple of astrobiology.

The surface temperature of the Earth is maintained by a greenhouse effect. The surface temperatures for the terrestrial planets in the absence of atmospheres are much more extreme: Venus 11°C (284 K), Earth −31°C (242 K) and Mars −78°C (195 K) [8].

Then, the nearly circular orbits of low inclination that we mostly take as a given, are much more difficult to achieve than originally supposed. Observations of exoplanets appear to favor orbits with high eccentricity. Thus the nearly circular orbits of planets in our system are perhaps exceptional. Estimates of the width of the zone around the Sun in which a habitable planet can reside in our solar system are quite narrow, ranging from about one-tenth to about one-half an AU around the orbit of the Earth (Figure 18). This requires that the orbits be nearly circular. Eccentric orbits might be unsuitable, although microbial organisms might tolerate long periods that were too cold or too hot, judging from our experience with "extremophiles".

When all else is right, the presence of other planets may influence the eccentricity of the orbit, in turn producing large variations in the flux received from the star and hence habitability. The width and location of the zone are not fixed but will vary with time as the star evolves. Many other factors, such as plate tectonics, a magnetic field, planetary rotation, tilt, presence of volatiles, type of star and position

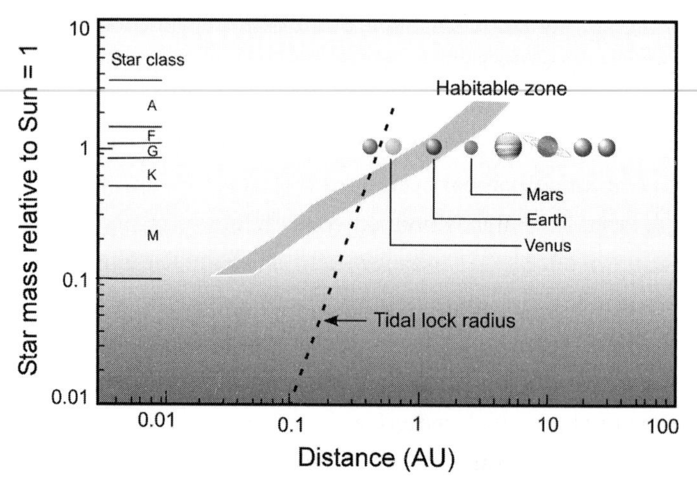

FIGURE 18 Habitable zones around planets

The "habitable zone" in which liquid water is stable on a planetary surface, plotted against the classes of stars, with our planets shown opposite the G class (not to scale). As the mass of the parent star decreases, the habitable zone moves closer to the star. Planets in such locations may be tidally locked into synchronous rotation, always presenting one face to the star in an eternal gaze. See also color plates section.

within the habitable zone are probably crucial. All these conditions needed for habitable zones seem to be highly restrictive. It is perhaps worth noting that occupying a habitable zone does not necessarily make a planet habitable.

Meanwhile classifiers have produced four classes of habitable planets: Class I like the Earth, Class II like Mars and Venus, Class III planets with subsurface oceans with silicate floors (Europa?) and Class IV, "ocean planets" and planets with liquid water between two ice layers. They conclude, unremarkably, that "considering all these four different classifications of habitable planetary bodies, it seems hard to imagine higher life forms as we know populating anything but a Class I habitable planet" [9].

One might deduce from this that we live in an up-market neighborhood and that as we deviate from planets exactly like the Earth, the prospect of habitability decreases rapidly. So in our system, the

Sun has the right metallicity, we are in the right place in the galaxy, the Sun is about the right size, has a long life, is stable, the Earth is in a circular orbit and has a suitable tilt.

This situation is often referred as being in the Goldilocks zone, after the children's story. Venus is too hot, Mars is too cold, but the Earth, like Baby Bear's porridge, is just right. We do indeed have a Goldilocks location. The bears have yet to arrive. Although the fable is often invoked, the Goldilocks fallacy may be briefly summarized: life, once started, will always adjust to the environment, as we see on Earth where an astonishing variety of forms have evolved to cope with every habitable niche. The porridge will always be "just right".

Stars more than 1.5 times the mass of the Sun (O, B and A classes) have lifetimes that are too short for life to evolve very far on suitable planets. Sirius, 25 times more luminous and twice the mass of our Sun, is expected to shine for only 1 billion years, too short in our experience for more than bacteria to emerge. Life seems most likely to survive around cooler F, G or warmer K stars that are stable over billions of years, although some of the larger F class have lifetimes of 2 billion years or less. If experience on the Earth is a guide, these periods are too short for evolution to proceed very far.

Red dwarf M-class stars, although they constitute over three-quarters of all stars, seem less suitable. The largest red dwarfs have only about 10% of the luminosity of the Sun. Most are much dimmer. Small stars like these dwarfs have habitable zones closer to the star. Two problems arise. The planet may become tidally locked into a synchronous orbit, presenting one face to the star, producing extreme temperatures on both sides of the planet. Thick atmospheres might even out the difference, but then might lead to a Venusian greenhouse (Figure 18).

Tidal heating may drive excessive volcanism that might produce unsatisfactory surface conditions for life. One cannot imagine even extremophiles existing on the Galilean satellite Io. Many other complexities may arise, such as a need to remove CO_2 and avoid a greenhouse developing. Life here uses light mainly in the visible spectrum, but dwarf stars emit little visible light. Gliese 581d is a possible example of these problems. It is the first planet discovered that is within the habitable zone and so might have an ocean, but it is probably tidally locked [10].

There is a certain irony that the "habitable zone" in our present solar system occupies a region in the inner nebula that was originally depleted in water, carbon and other volatile elements essential for life. The very elements critical for life are in short supply in the only regions where it might take root and flourish. If these elements, vital for life, come from accidental trapping in planetesimals or have to be re-imported from the asteroid belt or beyond, this makes the creation of a habitable planet and the emergence of life even more dependent on stochastic processes. So making a habitable planet depends on a complex set of factors, of which distance from the Sun is only one.

Other problems occur with the water content of planets in the habitable zone. Too much in a planet near the warmer inner edge of the zone might induce a greenhouse atmosphere like Venus. If it is located on the colder far side of the zone, the water might freeze, producing a permanent snowball planet. Desert planets might be safer to live on but suffer from too little water. All this supposes that the planet is in a circular orbit. But other planets might induce the orbit to become eccentric, so that the planet might move from being too hot to too cold, with problems for life. Clearly it is more difficult to achieve the right balance than Goldilocks experienced.

The origin and evolution of life

> "There is no master plan, no divine architect, no intelligent design. Nature endlessly experiments and we are simply one among the innumerable results" [11].

The whole question of the existence of Earth-like planets is inextricably intertwined with the debate over the existence of extraterrestrial life (which is usually assumed overtly or covertly to be intelligent). On this topic the biologists, familiar with the random course of evolution and contingency, have mostly been skeptical, while the physicists seeing determinism, have been less so.

The first essential for life as we know it is the presence of carbon and related elements. Life based on silicon is a favorite staple of science fiction writers. However, the chemistry of silicon is so distinct from that of carbon as to make it an unlikely candidate for being the principal component in building self-replicating molecules. The basic problem is that the silicon atom is much larger and lacks the ability to form the sort of complex bonds that carbon is able to form. Exhaling SiO_2 (quartz) rather than CO_2 might make breathing difficult, except at very high temperatures.

The substitution of arsenic for phosphorus in the DNA structure has also been reported, although it is not clear if the element is an integral part of the molecular structure. Perhaps the bacteria living at the bottom of an inhospitable lake may find themselves compelled to use what is available, rather than an element that they would prefer. But proof is lacking, no other scientists having substantiated the claims. This reminds one of the similar uproar over the unproven presence of fossil bacteria in a Martian meteorite.

The chemical elements needed for life, however, were not there "at the beginning" of the universe. Only hydrogen, which constitutes nearly three-quarters of the universe, helium (about a quarter) and a trace of lithium were produced in the Big Bang. The heavier elements, including carbon, oxygen, nitrogen, phosphorus, iron and other elements essential for life and also for making planets, were formed later by nuclear reactions inside stars and in supernova explosions. The nuclear furnaces in the stars have formed elements on a scale that would have astounded the medieval alchemists. As a star comes to the end of its life, it sheds mass or explodes. The newly formed elements are dispersed out in a patchy manner into the gas and dust of interstellar space. They provide the material from which new stars are born.

Clearly life could not begin until elements such as carbon, potassium and phosphorus became relatively abundant. It took several billion years and many generations of stars to produce enough of these elements. By the time that the solar system formed, the heavier elements, including the life-forming elements, formed between 1 and 2% of the gas and dust in the solar nebula. Thus, life is not some all-pervasive presence in the universe, but just another set of chemical compounds that has to await the appropriate conditions to form self-replicating molecules. Precursor organic molecules found both in space and in meteorites are common enough. It seems possible that life might arise anywhere given the right mix of chemistry and environment and so might well be common in the universe. The development of intelligence is another matter.

Fred Hoyle (1915–2001) and his co-workers interpreted some features in the spectra of Comet Halley as evidence for the presence of bacteria. Such an identification would have profound implications for the origin of life, since bacteria might arrive "ready-made" on the Earth from outer space. However, the fatal flaw in this entertaining concept is that the identification is non-unique. Many organic molecules show features that provide an excellent match to the spectral fingerprints of bacteria. It is not difficult to find look-alikes among the several million organic compounds that are known.

When Earth-mass planets are discovered, the next step will be to search their atmosphere for gases that might reveal the presence of life. These so-called "biomarkers" include oxygen, water and carbon dioxide that are characteristic of terrestrial life. Ozone plays a critical role in preserving life here in absorbing the more energetic ultra-violet photons from the Sun. It has long been regarded as a biomarker for oxygen, but its recent detection in the atmosphere of Venus casts doubt on its usefulness.

A reductionist view

This view expressed here is often referred to as "reductionist" and conflicts with the idea that life has a vital, mystical or ethereal component, an "élan vital" outside of physical reality, that is commonly referred to as the "soul" [12]. The concept of the soul as part of the mind, or the "ghost in the machine", seems to be a relatively recent concept, mostly due to the great French (or Dutch, for he spent his most productive years in Holland) philosopher, Rene Descartes.

Francis Crick (1916–2004) in his entertaining book *The Astonishing Hypothesis* presents a different view. It is that "you" are, in fact, no more than the end result of the behavior "of a vast number of nerve cells and their associated molecules" [13]. However, this seems to me, in the words of Thomas Jefferson (1743–1836), to be self-evident, rather than astonishing, and I continue this discussion on a thoroughly reductionist theme on the basis of scientifically observable and testable models.

Like gloves, complex organic compounds possess both left- and right-handed forms in equal proportions. If organic molecules had continued to arrive from outer space, both left and right-handed forms would have been delivered from meteorites. However, life on Earth has been curiously selective. The amino acids in living organisms are only "left-handed". In contrast, the nucleic acids are all "right-handed" molecules. Why is this so? No one knows, but it looks as though a single event or perhaps selective process was involved.

So the origin of life looks like a one-shot affair. Perhaps there was some obscure evolutionary advantage in selecting the left- or right-handed forms, or maybe it was the result of chance that persisted once the pattern was set. One is reminded of the "qwerty" arrangement of computer keyboards. This originated because of the need to separate frequently used letters to avoid jamming problems in early mechanical typewriters. The problem has long disappeared, but we are left with the fossil order, despite the existence of more efficient arrangements for keyboards.

The famous experiments of Stanley Miller (1930–2007) and Harold Urey in the 1950s produced abundant amino acids, possible precursors of life. Electrical sparks, simulating lightning, in a mixture of hydrogen, ammonia and methane that was thought to mimic an early terrestrial atmosphere containing those gases, achieved this feat. But we do not know the composition of the early atmosphere.

It is usually thought that the early atmosphere consisted of methane, water vapor, carbon dioxide and nitrogen, but there is little solid evidence.

Other models suppose that life arose in Darwin's "warm little pond". In such an environment, the complex organic compounds produced in the atmosphere by lightning, solar UV radiation, in addition to those brought in by meteorites, might achieve the requisite jump into a reproducing species. Clay minerals are often thought to provide a sort of template. There are innumerable other suggestions.

In those remote epochs, there is a scarcity of continental crust and only a few islands that might provide tidal pools. More serious is the devastating rain of asteroids and comets. Judging from the numbers and ages of craters on the Moon, impacts of massive bodies capable of forming basins hundreds or thousands of kilometers in diameter, were formed on the Earth. Such globally sterilizing events may have been concentrated at the time of the Late Heavy Bombardment, around 4000 million years ago. Perhaps the development of life was frustrated by this bombardment, which smashed up any early crust that had tried to form on the Earth, while the largest collisions removed any early atmosphere.

Deep under the oceans, life might arise in relative safety and later colonize the surface regions. Volcanic vents, often referred to as "black smokers", are common deep under the oceans at the mid-ocean ridges. They are a rich source of strange life forms. Among the most primitive organisms are sulfur-bearing bacteria and experiments have demonstrated that complex organic molecules can be formed under these conditions. These environments were widespread early in Archean times when most lavas were erupted under the oceans in the absence of appreciable land masses.

However life arose, it seems to have been well adapted from its beginning. Biology has built an astonishing variety of organisms using relatively simple modules, just as many organic chemical compounds and complex minerals have been constructed from a few basic building blocks. A good example are some 3 million current species of beetles, leading to the comment that Earth is "the planet of the beetles".

In this book, I have argued that the likelihood of developing copies of this planet, or of our solar system, are remote. However, it is of interest to observe the progress of evolution on this one example that we have available to us, and to enquire whether something resembling *Homo sapiens* might arise elsewhere. Evolution, however, has no necessary direction or plan to achieve this goal [14]. For the first 2 billion years, life on this planet was restricted to simple bacteria and archea. Although life arose within 1 billion years following the formation of the Earth, this does not tell us anything about the possibility of life arising on a suitable exoplanet. What it does tell us is that intelligent life took a long time to develop here and is the result of many chance and improbable events.

THE DEVELOPMENT OF INTELLIGENCE

> "If life exists elsewhere, what's to say it hasn't evolved into complex and intelligent beings?" [15].

In spite of the great flowering of life on this planet, intelligence seems to have developed very late. It occurs only among the vertebrates and there, rarely. Among the 24 orders of mammals, high intelligence seems to have arisen in only one, in humans. Why is this so? Clearly, high intelligence has little evolutionary advantage, for it has appeared once in tens of billions of attempts. As Ernst Mayr (1904–2005) the biologist, has pointed out [16], even the development of high intelligence may not lead to the ability to communicate with distant planets. Only one of the 20 or so civilizations that have arisen on Earth in the past 5000 years has developed the technology to communicate with other possible life forms elsewhere (Figure 19).

Although our species now needs rather more intelligence than is generally obvious, unfortunately the brain size of *Homo sapiens* seems to have stopped increasing around 20,000 years ago, the time of the Cro-Magnon cave painters. This evolutionary dead end seems

FIGURE 19 Former civilizations

Traces of former sophistication. The temple of Kukulcan at Chichen Izta in Yucatan, Mexico, built in 445 BCE, and used until 1204 CE. The temple (El Castillo) accurately predicts the Spring (March 20 or 21) and Fall (September 21 or 22) equinoxes, a tribute to Mayan science. The base width is 55 meters (180 feet) and the structure rises 24 meters (78 feet) surmounted by a 6 meter (20 feet) temple. The site of the K-T impact was also in Yucatan. The sacrificial well at this site is one of the sinkholes (cenotes) that mark the edge of the impact crater and which are probably related to subsidence along the crater rim. See also color plates section.

to have been reached because of the increasing mortality rates associated with childbirth, as discussed above. Two evolutionary trends seem to have reached a crossroads at that time. The need to walk on two legs requires that they be close together. But this restricts both the pelvic opening and dimensions of the birth canal and puts a limit on brain size. This seems to have been reached in Cro-Magnon Man (or Early Modern Humans) with a brain capacity of 1600 cm^3, compared with an average of about 1200 cm^3 for modern humans. We clearly are superb toolmakers. This is well illustrated by the high technology displayed for example, by color television. But our intellectual progress has lagged far behind so that this incredible invention is used to transmit a wasteland of trivia and violence.

SETI

> "Philosophical and theological beliefs are the main motivations for the belief in (S)ETI" [17].

Now that the marvellous technology exists, we have begun the search, driven by the belief that species on exoplanets may be at least as advanced in technology as are we. Such notions seem common enough to reappear continually in the current scientific literature, for example "as our society is in its technological infancy, another civilization capable of communicating over inter-stellar distances is likely to be enormously advanced compared to our own" [18]. In the absence of evidence, the use of double negatives in astrobiology is a common cautionary device. Carl Sagan often used them, as in his comments about Mars that "there is no reason to exclude from Mars organisms ranging in size from ants to polar bears" or, about the Mariner 9 photos of that planet, that there was "nothing in these observations that excludes biology" [19].

Given the right environment, life might arise given the appropriate chemistry. But the appearance of bacterial life, although interesting in itself, is not what people are looking for. The search for extraterrestrial intelligence (aka little green men) is really the issue. But in this search for extra-terrestrial intelligence (ETI) or SETI, several major obstacles appear. Apart from the distance and time questions, an Earth-like rather than an Earth-mass planet first has to form. The problems involved form a major theme of this book. Next, a similar geological history of planetary development needs to happen. An identical or closely identical history seems necessary. Then life has to begin and evolve in a closely similar pattern to life on Earth.

The development of a continental type crust that would form a platform on which a species could evolve seems required. Probably some mechanism like plate tectonics is required to build the crust. Then a successful species has to discover the chemical elements and the electromagnetic spectrum or develop interstellar travel.

A final problem, given that all the preceding ones have been satisfied, is that we will be able to understand an alien visitor or decipher any message. Never mind that we cannot communicate with the many millions of species (even many of our own) on our planet, although some elementary communication does occur with dogs, horses and perhaps a few other mammals. But hopefully the message will also be in modern English.

Many of us had difficulty in interpreting the message sent to outer space bolted to the struts of the Pioneer 10 and 11 spacecraft, launched in 1972 and 1973, now departing from our planetary system and expected to reach other stars in a few million years. Among other symbols, male and female nude figures of *Homo sapiens* were depicted on the plaque. A cartoon of the period showed a cocktail party with all in evening dress in progress in a museum on a distant planet. One guest contemplating the plaque commented that "Well, they are just the same as us except that they don't wear any clothes".

This comment is another way of stating that
We are looking for ourselves out there.

THE RISE OF SCIENCE

> "Though science is stronger today than when Galileo knelt before the Inquisition, it remains a minority habit of mind and its future is very much in doubt. Blind belief rules the millennial universe, dark as space itself" [20].

There is a general perception that science has arisen as a natural, indeed inevitable intellectual development, the laws of physics, for example, waiting for their Newton to arise. Perhaps the best evidence for this point of view is that few new scientific concepts occur truly independently. Parallel discoveries are very common, often leading to disputes over priority. The famous example is the development of calculus by both Newton and Leibniz. But these and similar discoveries appeared within a similar cultural community.

Even now, scientific thinking seems alien to many people, who prefer the comforts of mythology or religion. The popularity of explanations of natural phenomena based on the notions of primitive desert tribes, magic or astrology (there are currently more astrologers than astronomers) is often lamented. Such views that illustrate the limitations of the human intellect are highly suited to primitive hunter-gatherer societies. Objective and analytical thinking is not restricted by race or gender, but, like a delicate flower, requires a favorable environment in which to flourish. It is rather readily overwhelmed in faith-based or otherwise restrictive cultures. Curiosity is a common human trait, but is readily stifled by simplistic explanations, such as "God made everything" and its offshoots, "creationism" and "intelligent design". Even today science is struggling in many societies. The New Library in Alexandria, hopefully a replacement for the ancient Great Museum and Library, remains closed at the time of writing.

As Edward Gibbon remarked in his *Decline and Fall of the Roman Empire* (that monument to the Enlightenment, published in 1776) "the practice of superstition is so congenial to the multitude that, if they are forcibly awakened, they still regret the loss of their pleasing vision. Their love of the marvellous and supernatural, their curiosity with regard to future events, and their strong propensity to extend their hopes and fears beyond the limits of the visible world were the principal causes which favoured (sic) the establishment of Polytheism. So urgent on the vulgar is the necessity of believing, that the fall of any system of mythology will most probably be succeeded by the introduction of some other mode of superstition" [21].

It is against such attitudes that scientific modes of thinking are struggling. Superstition, religion, magic and astrology remain curiously popular in current societies that owe most of their prosperity to the ideas developed in the Renaissance and the eighteenth century Enlightenment in Europe.

So how did the concept of science arise in a species whose brain size ceased to grow at least 20,000 years ago? Historical research suggests that the concept arose in Ancient Greece in a particularly favorable environment and later took root in Western Europe, following the Renaissance [22]. In this view, the rise of scientific modes of thinking arose as a consequence of a peculiar combination of circumstances. These included the ability to think objectively, a weak religion and an open society that allowed discussion on any topic. Although there were many gifted individuals in Chinese, Indian and Ottoman societies, curiosity is easily repressed. Objective thinking seems particularly difficult and requires a society open to unrestricted debate.

Ancient Greece was a maritime nation, its citizens ranging widely across the Mediterranean. The many cities and ports that were visited commonly had distinct religions, each convinced that their particular cult held universal truths. Exposure by the Greeks to such attitudes encouraged a questioning attitude that was not inhibited by the multitude of gods in Greek religions. Dogmatic attitudes, so common in most societies, were weak and argument and debate was common. Greek culture was spread widely by the conquests of Alexander, extending as far as India. These factors persisted for about 1000 years, many of the intellectual developments taking place in Alexandria. This development of science contrasts for example, with the lack of objective thinking in societies such as ancient Israel, dominated by dogmatic belief in one all-powerful god.

The social circumstances in Greece or Alexandria were not readily repeated elsewhere. Although many remarkable technical developments occurred in China, Mughal India and the Ottoman Empire, none of these societies seems to have developed objective modes of thinking. All had received telescopes and knew of Galileo's observations within a decade of their discovery. Chinese society seems to have been dominated by bureaucracy and Indian, by mysticism. Both societies built sophisticated observatories. Good examples are at Beijing, built in 1422 and the Jantar Mantar Observatory at Jaipur, which dates from about 1730, curiously without telescopes that had already been known in India for 100 years [22]. Few fundamental advances in knowledge occurred compared to those that resulted from Tycho Brahe's observations in Denmark or Galileo's in Italy. Although telescopes and Western scientific literature were available in Chinese, Indian and Ottoman societies, few new concepts seem to have appeared [22].

Astronomy in these societies was mostly utilized for making astrological predictions and became the perogative of a bureaucracy or priesthood. Fortunately, the great intellectual heritage, before it decayed in eastern societies, managed to be transmitted to Western Europe. Here it took root and flourished, because of the unique development of legally autonomous institutions such as universities, where independent enquiry could proceed, less impeded by the religious or social restrictions common in other societies [22].

Counter arguments are that the discovery of gravity, Newton's laws of motion, the benzene ring, the electromagnetic spectrum or the chemical elements was inevitable, just things waiting to be discovered, if not by Newton, then someone in another milieu, perhaps in Patagonia or Papua. But it does not seem inevitable that modern science had to emerge. Nothing comparable has happened in the 20 or so civilizations that have arisen in the past few thousand years, although many significant technical advances have occurred. Only one of the 20 civilizations, one that had highly developed skills in astronomy (Figure 19), that have arisen on Earth in the past 5000 years has developed the technology with the potential to communicate with other possible life-forms elsewhere.

Although the discussion here may be considered by some as unduly Eurocentric, history, like geology, is about what happened, not what should have happened.

There seem to be too many contingent factors involved. One of the popular hobbies is to consider the "what ifs" of history, the counterfactual stories of what might have happened. Suppose that the

Greeks had lost the Battle of Marathon, that the Moors had overrun Europe in the eighth century or that the Counter-Reformation had been successful, Western civilization would then have taken a different course and science might never have developed. There seems as little determinism in human history as in evolution. Without that accident of plate tectonics in East Africa, we would not be here.

There are many popular misconceptions about science and its practitioners. The field is populated by a wide spectrum of humans, from classifiers of the infinitesimal to mystics embracing holistic views. Common sense and imagination are useful attributes. An ability to see interconnections among remote topics would seem to mark out some distinctive feature of the mind, as well as an ability to detect nonsense. The writings of the Nobel laureate and zoologist Peter Medawar or of the Cambridge philosopher Simon Blackburn provide examples of the former while the eccentric philosophy of Teilhard de Chardin represents an extreme case of an egocentric view of our position in the cosmos [23].

Opinions of amateurs on many matters are now widespread, assisted by the worldwide computer networks. But they suffer from a common defect, a lack of self-criticism that is central to scientific investigation. Topics as diverse as water divining, climate change, cosmology, earthquake prediction and evolution are popular topics for uninformed opinions. As a pastor in Dover, Pennsylvania lamented "We are being dictated to by the educated folk". Science seems very much under attack not only from creationists in Western societies but from extremist groups everywhere. *Homo sapiens* seems unwilling to accept that he is responsible for changing the climate of this planet or for the current "sixth" major extinction that is now upon us, nor for that matter, his earlier destruction of the Pleistocene megafauna.

One is reminded of what has been called "the failure of nerve" in the ancient world. Following the scientific advances of ancient Greece and the major progress at the Museum and Library at Alexandria, there followed a retreat into the comfort of myths. The development of science can only flourish in open secular societies. Well-funded universities in dogmatic societies with dominating religions seem doomed to failure.

Science and religion

> "Reconciling science and religion remains a popular project, especially among academics reaching the end of their careers" [24].

One approaches the topic of faith with caution. It is not a subject really open for discussion in polite society, where people are allowed to believe in any set of fantasies of which the 39 Articles (dating from 1563) of the Church of England form one of the more sober examples. This reliance on faith rather than reason merely illustrates the intellectual limitations of our species. However, the subject of faith needs to be raised here as, historically, there has always been a strong religious bias in discussions about the plurality of worlds, angels, aliens or UFOs. Indeed it has frequently been suggested that belief in extraterrestrials constitutes a modern myth, cult or religion. Certainly the seekers after Earth-like planets or the adherents of SETI often display a zeal not readily distinguished from that common among the "true believers" in religious sects.

Religion and science however, remain totally incompatible. Religion is based on faith and science on facts and reason. Humans and their affairs occupy both a thin envelope around a minor planet and a miniscule fraction of the time that the universe has been in existence. The universe is indifferent to the presence or absence of *Homo sapiens*. No profound knowledge of philosophy or of particle physics is required to reach this conclusion. A simple consideration of time and place, of space and time, the scale of things, suffices.

Conventional religious beliefs seem merely to reflect the current limited stage of evolution of the brain of *Homo sapiens*, so that the decline of one set of beliefs will be replaced by another, equally bizarre. Estimates of the number of gods that have been worshipped range from hundreds to millions, many assigned to specific tasks; for example, Thor's specialty was thunderstorms. More recent religions have tended to place all the responsibility on one individual, leaving thunderstorms to the meteorologists.

Religions seem to be based on the egotistical attitude of *Homo sapiens* toward his place in the universe. Whether this is achieved by the discovery of stone or gold tablets, the passage of a comet, resurrections, virgin births, the arrival of aliens in flying saucers or other unsubstantiated events outside rational enquiry, little seems necessary to bring about the establishment of cults and religions. These have numbered about 10,000, distinguished only by the number of adherents they attract, the utter belief of their followers that their version is the true one and by their ability to acquire legal status and hence exemption from tax.

One of the more eloquent spokespersons for faith, St Augustine (354–430 CE), ordained that, as God had created everything, "It was not necessary to probe into the nature of things" as the Greeks had done. However, only one set of physical laws, discovered by patient inquiry, enables an aircraft to fly. Religious believers seem to be convinced enough of these principles of physics and the laws of thermodynamics that allow jet engines to operate, as they travel about in aircraft. This acknowledgement reveals an interesting human trait. Most people accept without question that the Earth is round and orbits the Sun, although both notions seem to contradict common observation. Atoms and cosmology, both difficult to discover or observe, are widely accepted. But the demonstrated facts of evolution seem too uncomfortable for the human ego.

Physicists such as Paul Davies (b. 1946) want to see meaning in the universe. His ringing statement that "I cannot believe that our existence in this universe is a mere quirk of fate, an accident of history, an incidental blip in the great cosmic drama. This can be no trivial detail, no minor byproduct of mindless, purposeless forces. We are truly meant to be here" [25] sums up his philosophy. Davies notes that some people, such as the biologist Jacques Monod (1910–1966), express a differing view. Monod has made the famous comment that "nature is objective" and that "man knows at last that he is alone in the unfeeling immensity of the universe, out of which he has emerged only by chance. Neither his destiny nor his duty have been written down" [26].

Faith seems dangerous. Faith makes frequent appeals to spirituality, mystical and transcendent experiences. But these have little to do with reality and are merely one consequence of our large brain and our well-developed capacity for imagination, self-delusion and hallucination. It is ironic to see modern sophisticated societies, based on the scientific discoveries of the past few centuries and facing a probable catastrophic future, reverting to the folklore of primitive desert tribes. One hopes that science which is a social activity embedded within societies will survive the current wave of religiosity in the United States.

It seems to be better to stand up and face the objective evidence for what it is, rather than behaving like the mythical ostrich and burying one's head in the sand. The knowledge that we are probably alone in the universe, that conscious intelligence has arisen accidentally and we are its only keeper should stimulate us to behave more responsibly. Our outstanding achievement has been to discover where we are, how we got here and what is going on. Where else in the universe might there be someone who knows about the exotic chemistry of the rare earth elements, DNA, or that the galaxies are flying apart? Religious folk sometimes take comfort in the fact that we may be alone, seeing this as evidence of special creation. But this is a delusion. Both our useful planet and *Homo sapiens* are here because of a long series of chance events and accidents.

The message of this book is clear and unequivocal: many chance events have happened during the development of the universe over the past 14 billion years. The bewildering variety of exoplanets lends support to this conclusion. Superimposed on these chance events from the physical world are those of biological evolution that has managed to produce one highly intelligent species out of tens of billions of attempts over the past 4 billion years.

Design?

> "If we were to see the hand of a designer anywhere, it would be in the fundamental principles, the laws of nature. But, contrary to some assertions, they appear to be utterly impersonal and without any special role for life" [27].

The proponents of design among animals, having been defeated by the observations of Charles Darwin amongst many other biologists, have now retreated to the debatable comforts of particle physics. The craftsman who displayed his handiwork among the many millions of beetles has now been replaced by a remote particle physicist who tweaked the constants of physics before the Big Bang, so that *Homo sapiens* could develop 14 billion years later.

The list of physical constants that make our existence possible is certainly impressive: Among many examples, there are some obvious properties of substances, such as the curious chemistry of water and the density of ice. Most solids are denser than their liquid forms. But ice floats on water. If it didn't, it would sink to the bottom of lakes and oceans, and never thaw, making life, as we understand it, very difficult.

The masses of protons and neutrons are just about right to enable hydrogen to form. Then there is the famous example of the difficulty of forming the element, carbon, during element synthesis in stars. The existence of a metastable state in the binding energy of the carbon nucleus enables it to exist just long enough, not only to preserve the element carbon, but also to enable the chain to proceed to make heavier elements, such as oxygen. But the constants are not as finely tuned as they might appear and the value for carbon could be 20% higher, as Stephen Weinberg has noted [27]. Others dismiss the correlations among large numbers as mere coincidences more suited to numerology than to physics.

Perhaps these coincidences are just "brute facts" with no more significance than the fact that the Sun and the Moon have almost exactly the same apparent size in the sky. Is that some kind of cosmic joke, put there to fool us, or is it merely a remarkable coincidence?

Possibly most of the fundamental constants evolved as a consequence of the Big Bang, just as the stability of the chemical elements is fixed by the ratio of protons to neutrons in the atomic nucleus. If the constants were different, or if no carbon had been formed, some separate universe or form of life might have arisen, just as the Periodic Table would contain some other interesting elements if the binding energies of protons and neutrons were different. It is of course true that all of these properties are essential to our existence, just as the correct value of pi (π) is essential to the construction of a wheel [28]. Maybe it is just the way things evolved, totally unrelated to our existence.

The anthropic principle

> "There is a sociology as well as a logic to science. Some bad scientific ideas can spread widely, at least for a while" [29].

Science has dethroned *Homo sapiens* from his central position and relegated the species to a distant corner of the universe. It was bad enough when Nicolaus Copernicus put the Sun, rather than the Earth, at the center of the world. Worse was to follow. Geologists discovered the unfathomable "dark backward and abysm of time" [30] of an extent unimaginable on human timescales (Figure 16). Next on the scene was Charles Darwin who placed *Homo sapiens* among the animals. Then Edwin Hubble (1889–1953) found that our Milky Way galaxy, whose edge-on view is at least visible to us, was only one speck amongst a great host. Lately we have discovered that not only are the galaxies receding, but they are accelerating away from us. Now a great host of exoplanets have been discovered, but nothing looks like our familiar solar system and our friendly Earth.

This new knowledge from science seems to have damaged our egos, perhaps deeper than we realize. A reaction to these views is found, amongst many others, in Sagan's "principle of mediocrity", that the universe is mostly similar but manages to ignore the fact that most stars are tiny red dwarfs. Surely we were meant to be the Lords of Creation, occupying a central place in the cosmos, rather than the inhabitants of some obscure, even if interesting, backyard.

However, the human ego is stronger even than that of the household cat, *Felis domesticus*. Now our self-interest has attempted to bring our existence back to a place of central importance in the universe. Amongst the more respectable attempts is the anthropic principle [31] and it finds comfort for us among the constants of physics. It is the latest version of design arguments that go back to Aristotle.

There are variations of the anthropic principle, usually referred to, like tea or alcoholic drinks, as strong or weak. Like theological dogma, these principles are a bit difficult to pin down with any precision. The strong variety seems to say that the universe must have the properties to enable life to arise and so life must arise at some stage in the universe. This would imply a conscious act by a creator, and so be beyond scientific investigation. It should be noted, however, that the principle does not specify that the life that arises as a consequence must be intelligent, although no doubt the authors of the idea intended this. This oversight recalls the fate of Tithonus, the mortal lover of Eos, Goddess of the Dawn in Greek mythology. She persuaded Zeus to make him immortal, but forgot to ask the god to grant him eternal youth. Eventually Tithonus became helpless with old age, but he talked incessantly and so was shut away, perhaps a cautionary tale for philosophers.

The weak variety of the anthropic principle is a little more flexible and says that the physical properties of the universe have taken on those values to enable human beings to exist and measure them. This merely states that we live in a universe that we can observe, and seems to be a statement of the obvious.

It is ironic that we now see apparent design in physics, just as William Paley saw the hand of the designer in biology, 200 years ago. Darwin explained the reasons for biology. Are the physicists still waiting for their Darwin to explain the reasons behind the apparent evidence of a designer who set up the physical constants for us? It seems possible that eventually the "large numbers" coincidence, the ratio of photons to baryons and so forth, will be able to be calculated from some "Grand Unified Theory of Everything". Then these curious coincidences will not need to be accounted for by the anthropic principle, any more than we now need the god Thor to account for thunderstorms.

The anthropic principle suffers from that fatal defect of scientific hypotheses of being untestable. For this reason, it is a philosophical curiosity somewhat like the Drake Equation or the Gaia Hypothesis. It is interesting that the anthropic principle refers exclusively to *Homo sapiens* and the purpose of his being here. But the dinosaurs, who dominated the Earth for 160 million years, had a better case to be considered lords of creation. Unaware of the accidental collision waiting to remove them, the dinosaurs would have had good reason to invent a "Reptilomorphic Principle" to explain why they had ruled the Earth for so long. The beetles, all 3 million species, without counting in their multitude of ancestors, have an even stronger case for regarding this planet as designed for them. And what about the billions of foraminifera in the oceans, or anything else on this planet, none of which would exist if the physical constants were different?

The anthropic principle looks like another desperate attempt to put *Homo sapiens* comfortably back on center stage, a view compatible with the authors of various religious texts. However, a retreat into medieval modes of thought on our over-crowded planet would bring catastrophe.

CHANCE EVENTS: THE ANTIDOTE TO THE ANTHROPIC PRINCIPLE

> "The 'misanthropic principle' is the observation that . . . the environments and situations necessary for intelligence to develop are extraordinarily rare" [32].

Many chance events in the physical world directly affected the origin and evolution of life and our existence on Earth. The list is impressive. I begin with the size of the fragment that broke away from the molecular cloud. If the fragment had been bigger, or spinning more rapidly, it might have spun out to a dumbbell and made two stars. Then it took fine timing to form Jupiter before the gas was all swept away. Without the shield of that giant planet, we would suffer a continual bombardment of comets. If more Jupiters had formed, we might have a system with only two planets, one closely orbiting the Sun every few days and the other far away in a wildly eccentric orbit.

If the Earth was a little smaller or drier, the basaltic lavas could not recycle back into the mantle. The 500 ppm of water on the Earth was accidentally contributed by icy planetesimals but it made plate tectonics possible. Without water, we would have no granite, no continents to stand on, few ore deposits and no advanced technology. The continents enabled the final stages of land-based evolution to proceed above sea level and so enable this narrative to be written. The barren basaltic plains common throughout the other solid planets are less inviting.

The rotation rate of the Earth, which we take for granted, is a probable consequence of the great collision that formed the Moon. Without that accident, the Earth might have resembled Venus, and rotated slowly backwards. The tilt of the Earth that provides the seasons, celebrated by so many composers and painters, is the result of the same accident that also removed any thick primitive atmosphere.

But perhaps the most dramatic accident of all was the massive dinosaur-killing impact that closed the Age of Reptiles. Any values that the physical constants had were irrelevant to the event. It seems likely that if the asteroid had missed, the descendants of the dinosaurs would now dominate the planet. Our ancestors would never have walked on the plains of Africa and I would not be sitting at a word processor writing this account.

DEEP TIME

> "Believers in extraterrestrial intelligence have tended to lack what might be termed 'a sense of history'" [33].

It is certainly very difficult for *Homo sapiens*, accustomed to periods of 50, 100 or 1000 years, to understand the stupendous extent of geological time. The infinity of space seems easier to understand than the concept of deep time. Perhaps this is due to the concept of the light year. Every schoolchild knows that it takes 8 minutes for light to reach us from the Sun and 4 years for light to travel from the nearest star.

James Hutton (1726–1797) was among the first to demonstrate the immensity of geological time, particularly well illustrated by the tens of millions of years required to account for the formation of angular unconformities in the rock record (Figure 16). Geologists casually refer to periods of millions of years, but it is difficult to envisage the passage of even 1 million years. Here is one example that illustrates the immensity of time.

Limestone (commonly referred to as The Chalk) that was deposited during the Cretaceous Period forms the White Cliffs of Dover, famous in song, literature and legend. Beachy Head, a cliff made entirely of chalk, on the south coast of England, rises 162 meters or about 520 feet (Figure 20). Another example is the great white Shakespeare Cliff, near Dover, which is over 100 meters (330 feet) tall. The total thickness of The Chalk, now much eroded, was originally about 300 meters (more than 1000 feet) thick. The chalk was deposited by a rain of coccolith (nanoplankton) fragments a few microns in size on to the Cretaceous sea floor that occupied much of what is now Western Europe. This entire 300-meter thickness of The Chalk was deposited in about 30 million years, the dates being well bracketed by geology. Of course there were some places or times in which it was deposited rapidly, in others, slowly or not at all. But the overall accumulation rate was at an imperceptible rate of 1 millimeter per century.

Within human time frames, this would mean that during the 200 years since Mozart composed his unique music, or since the Battle of Trafalgar, 2 millimeters would have accumulated; only 4 millimeters from Shakespeare's time, 2 centimeters since the beginning of Christianity, 6 centimeters in Archbishop Ussher's 6000 years and 10 centimeters since the ice sheets melted.

FIGURE 20 Deep Time

Deep Time. Beachy Head, a cliff made entirely of chalk, on the south coast of England, rises 162 meters or about 520 feet. The chalk was deposited by a rain of coccolith (nanoplankton) fragments a few microns in size, on to the Cretaceous sea floor that occupied much of what is now Western Europe. The overall accumulation rate was at a rate of 1 millimeter per century, so that the chalk exposed in this cliff represents the passage of over 16 million years. See also color plates section.

Even the stupendous span of 30 million years needed to deposit The Chalk is less than 1% of the time that the solar system has existed, which in turn is about one-third of the time since the Big Bang. Human existence is trivial on this scale. It seems well known that we are lost in the immensity of space, occupying a minute speck in the universe. But *Homo sapiens* is also marooned in deep time.

THE SPECTRE OF COINCIDENCE

Searchers seeking extra-terrestrial intelligence understand the "tyranny of distance" well enough but seem less aware of the "tyranny of time". The problems of communicating or travelling over the vast expanses of interstellar space are a familiar topic, but in looking for intelligent life elsewhere, an equally serious problem is that of co-incidence. The window of time on the ETI planet has to correspond exactly with ours. Biological evolution as well as planetary development has to coincide. A trivial few hundred years either way would ruin the chances of communicating.

Not only is it necessary for all of the accidents that resulted in our appearance to occur on another suitable planet to produce some equivalent of *Homo sapiens*, but we also need them to discover within the past 150 years, either the principles of electro-magnetic radiation or how to traverse vast interstellar spaces. It is of little use if the alien spacecraft or signal arrived here during the Cretaceous, when the chalk now in the White Cliffs of Dover was accumulating on the sea floor at a rate of 1 millimetre per century and *Tyrannosaurus rex* roamed the outwash plains of the Hell Creek Formation of Montana and the Dakotas. Equally unsatisfactory would be contact with reptiles in a Triassic desert, giant dragonflies in a Carboniferous swamp, or with bacteria during the 3 billion years of the Precambrian. Contact has to happen on "this bank and shoal of time" [34].

Among the coincidences needed for communication with an intelligent species on an exoplanet, are:

- An Earth-mass planet with a similar composition, including a splash of water and possibly with a stabilizing moon, forms in the habitable zone around a suitable star.
- The planet follows an analogous geological history to the Earth. Near enough is not good enough, as Venus tells us.
- Plate tectonics begins and continents form.
- Life begins and evolves, surviving catastrophes.
- A suitable atmosphere arises.
- An intelligent species evolves.
- The species develops an advanced civilization, discovers the chemical elements and the electromagnetic spectrum. The species develops either interstellar means of travel or communication using some part of the electro-magnetic spectrum.
- The planet is close enough to enable communication on reasonable timescales: 1250 light years is one popular estimate.

- All of these developments must coincide exactly in time with the recent arrival of *Homo sapiens* on the Earth and with the scientific developments of the past 150 years [35].

When the remote chances of developing a habitable planet discussed in this book are added to the chances of life beginning and developing both high intelligence and a technically advanced civilization at precisely the same time as us, the chances of finding "little green men" elsewhere in the universe decline to zero. This might not be a conclusion that one wishes for, but in the nature of science, this follows from the data.

Appendices

SOURCES

This book is derived from many discussions and sources, the first six books being particularly useful:

Apai, D. and Lauretta, D. *Protoplanetary Dust*, Cambridge University Press, 396 pp., 2010.

de Pater, I. and Lissauer, J. J. *Planetary Sciences*, second edition, Cambridge University Press, 647 pp., 2009.

Lodders, K. and Fegley, B. *Chemistry of the Solar System*, RSC Publishing, Cambridge, UK, 476 pp., 2011.

McFadden, L. *et al.* (Editors). *Encyclopedia of the Solar System*, second edition, Academic Press, 935 pp., 2006.

Perryman, M. *The Exoplanet Handbook*, Cambridge University Press, 410 pp., 2011, which contains over 3200 references.

Seager, S. (Editor) *Exoplanets*, Arizona University Press, 526 pp., 2010.

Taylor, S. R., *Solar System Evolution: A New Perspective*, Cambridge University Press, 460 pp., 2001.

Taylor, S. R. On the difficulties of making Earth-like planets (Leonard Medal address). *Meteoritics and Planetary Science*, **34**, 317–329, 1999.

Taylor, S. R. and McLennan, S. M. *Planetary Crusts: Their Composition, Origin and Evolution*, Cambridge University Press, 378 pp., 2009.

The series of books on planetary and space science published by the University of Arizona Press are especially relevant to the theme of this book, in particular the following volumes:

Asteroids III (Editors: W. Bottke *et al.*) 2002.
Jupiter (Editor: Fran Bagenal) 2007.
Meteorites and the Early Solar System II (Editors: D. Lauretta and H. Y. MacSween) 2006.
Protostars and Planets V (Editors: B. Reipurth *et al.*) 2007.

I have also surveyed much of the scientific literature dealing with exoplanets through 2011 in journals such as the *Astrophysical Journal, Icarus, Nature* and *Science*. This flood of papers resembles a mountain torrent, with close to 10,000 papers being published on the topic of exoplanets since 1995. The difficulty of extracting information resembles that of trying to get a drink from a fire hose. I do not include references to URLs and conference abstracts, as they are too ephemeral for incorporation here.

In a scientific book, it is normal to make detailed references to the sources of the information that have been used. Here I have tried a different approach, writing more in the style of an essay, with only a few formal references. These are identified in the text by number in each section, thus [6], and are listed by section at the end of the book.

THE GEOLOGICAL TIMESCALE ON THE EARTH

There have been many attempts to tabulate the extent of geological time or to depict the geological timescale in graphical representations. Innumerable representations and variations exist and are widely available. The compilation by the Geological Society of America is recommended for those seeking further detail. Here I do not attempt yet another table or diagram, but instead give a list of the divisions of time that have been mentioned in the text. It is mostly arranged from youngest to oldest, as is customary. I have added a few comments where appropriate and have generally continued to employ common usage with apologies to the International Stratigraphic Commission. I have also tended mostly to ignore terms like eons, era, systems and the like that are of interest mainly to specialists.

Anthropocene

This is the latest informal subdivision of geological time that has been proposed, on account of the influence of *Homo sapiens* on the outer environment of the Earth. A popular date for its beginning is 1850 CE, the start of the "industrial revolution". Whether this term, or *Homo sapiens* itself will survive, is outside the scope of this work.

Quaternary

This period began 2.6 million years ago, marked by the ice ages and the evolution of *Homo*. It is subdivided into the earlier **Pleistocene** that ended 12,000 years ago and the **Holocene** or **Recent** which extends from the melting of the last ice sheets to the present day, or to 1850 CE, according to taste.

The **Tertiary** includes the time from the dinosaur extinction 65 million years ago to the Quaternary. But readers should be aware that the term **Cenozoic** is officially used to include all of the above.

Mesozoic or The Age of Reptiles

This is divided into the three familiar periods of **Triassic** that began at 251 million years, the **Jurassic**, dating from 202 million years ago and the **Cretaceous** that began 146 million years ago. The Cretaceous ended dramatically at the K-T Boundary, with the impact of an asteroid that ended the "Age of Reptiles". That event extinguished the dinosaurs, plesiosaurs, mosasaurs, ammonites and much else 65 million years ago, paving the way for the rise of mammals and us.

Paleozoic

This extends from the beginning of the detailed fossil record at the "Cambrian Explosion" dated at 542 million years ago and ends with the "Great Killing" that marks the close of the Permian Period, 251 million years ago. The Periods that constitute the Paleozoic, in descending order of age are:

The **Permian** that began 299 million years ago.
The **Carboniferous**, started at 359 million years ago, the "Age of Coal", divided in the United States into Mississippian and Pennsylvanian.
The **Devonian**, the "Age of Fishes" that began 416 million years ago and closed with another major extinction.
The **Silurian**, began 444 million years ago.
The **Ordovician**, that witnessed a major extinction and began 488 million years ago.
The **Cambrian** that began with the explosion of shelly fossils, dated at 542 million years ago.

Precambrian Era

This stupendous extent of time started about 4000 million years ago and extended up to the beginning of the Cambrian Period. It is divided into the **Archean** and the **Proterozoic**, with a boundary set conventionally at 2500 million years ago between them. Many sophisticated subdivisions of the Archean and the Proterozoic have been suggested. All need to be based on radiometric dating, in the general absence of fossils, except in the Ediacaran Period in the Late Proterozoic.

The Archean was marked by the emergence of life, the slow development of continents and the appearance of plate tectonics in the Late Archean. Shortly thereafter in the Proterozoic, complex cells and an oxygen-rich atmosphere developed. Glaciations and probable "Snowball Earths" (the **Cryogenian**, 850–635 million years ago) and the exotic fossils of the **Ediacaran** (635–542 million years ago) mark the close of the Proterozoic and the beginning of the Cambrian explosion.

Hadean

This term is used to describe the time before 4000 million years ago and the appearance of an identifiable rock record. A few zircon crystals have survived, the oldest being dated at 4360 million years ago.

The next oldest reliable datable event is around 4460 million years ago, given by the crystallization of the crust of the Moon. This means that the Earth itself was mostly formed by that time. About 100 million years had elapsed since the formation of the earliest dated solids in the solar nebula at 4567 million years ago, the easily remembered date of T_{zero}.

THE GEOLOGICAL TIMESCALE ON MARS

Mars is the only other rocky planet for which we have some idea of the geological record. Purists have argued that geology refers strictly to the Earth, but the etymological derivation is suspect. Here the terms geology and geological are used for all the terrestrial planets as it makes for simpler sentences.

The geological record of Mars is divided into three major epochs: **Noachian, Hesperian** and **Amazonian**. There are only a few well-dated events, including some from the meteorites from Mars. Otherwise, the sequence is based on crater counts, but the craters are often obscured. Most of the geological record is compressed into the first billion years of the planet's history. The boundary between the oldest Noachian and Hesperian epochs is between 4100 and 3700 million years ago, so the Noachian of Mars is close in age to the enigmatic Hadean on Earth. The Hesperian/Amazonian boundary is around 3000 million years ago, within wide errors, so that the Amazonian on Mars is equivalent to all time on Earth since the Late Archean.

THE GEOLOGICAL TIMESCALE ON THE MOON

The Moon, somewhat like Mars, has a much simpler geological record than the Earth. The sequence was based originally on the cratering record, but the Apollo Missions provided many radiometric dates. The oldest is the **Pre-Nectarian** that extends from the formation of the Moon to 3920 million years ago. Like the Noachian on Mars, this covers most of Hadean time on Earth. The **Nectarian** covers the period of the Late Heavy Bombardment down to the formation of the **Imbrium** basin at about 3850 million years ago.

The **Imbrium** is divided in two. The "Lower" (or older) Imbrium extends for 100 million years to the formation of the Orientale impact basin (Figure 10) at 3750 million years ago. The "Upper" (or younger) Imbrium extends from the formation of the Orientale basin down to the age of the youngest dated lavas at 3200 million years ago.

The **Eratosthenian** that extends down to the formation of the crater Copernicus, about 800 million years ago, follows the Imbrium. The **Copernican** covers the time from the formation of that crater down to the present.

Notes

The following are the sources of quotations in the text and various notes on topics mentioned as asides to the main topic of the book.

PREFACE

1. I use the scientific term *Homo sapiens* for human beings throughout the book, thus avoiding the politically incorrect term, mankind and its ugly politically correct alternative, humankind. I employ *Homo sapiens* in a politically incorrect masculine usage as this makes for simpler sentences.
2. Seager, S. *Exoplanets*, Cambridge University Press, p. 441, 2009.

I PROLOGUE

1. Chambers, J. in *Exoplanets* (Ed. S. Seager), Cambridge University Press, p. 197, 2010.
2. Schneider, J. *et al.* Defining and cataloging exoplanets, *AA* Vol. 532, p. A79, 2010. As a provisional conclusion I therefore chose at this point to include all objects with a mass below 25 Mjup in orbit around a star as planets.
3. de Pater, I. and Lissauer, J. J. *Planetary Science,* Cambridge University Press, p. 538, 2010.
4. Cromer, A. *Uncommon Sense: The Heretical Nature of Science,* Oxford University Press, 140 pp., 1993; see also Huff, T. E. *Intellectual Curiosity and the Scientific Revolution,* Cambridge University Press, 368 pp., 2011. Huff gives the alternative spelling of Lipperhey for Lippershey, the inventor of the telescope.
5. Freeman, C. *The Closing of the Western Mind: The Rise of Faith and the Fall of Reason,* Heinemann, London, 472 pp., 2002.
6. Hughes, D. W. The Star of Bethlehem, Nature Vol. 264, pp. 513–517, 1976; Vol. **268**, pp. 565–567, 1977.
7. Laplace, P. S. *The System of the World* (English trans. J. Pond), R. Phillips, London, p. 123, 1809.

8. Boorstin, D. J. *The Discoverers*, Vintage Books, New York, p. 296, 1985.
9. Personal communication, Prof. Maria Firneis, Astronomy Dept., University of Vienna, 1996.
10. An extraordinarily useful source is Crowe, M. J. *The Extra-Terrestrial Life Debate*. University of Notre Dame Press, 554 pp., 2008.
11. Tipler, F. J. *QJRAS* Vol. **22**, p. 136, 1981.
12. Dick, O. L. *Aubrey's Brief Lives*, Secker and Warburg, London, p. 94, 1958.
13. Brush, S. L. *A History of Modern Planetary Physics*, Vol. 1, Cambridge University Press, p. 20, 1996, gives an account of this celebrated exchange between Laplace and Napoleon. See also Gillispie, C. C. *Pierre-Simon Laplace (1749–1827) A Life in Exact Science*, Princeton University Press, Princeton, NJ, 322 pp., 1988 for an account of the life of Laplace.
14. An excellent account of the origin of the Bible, based on archeological evidence, can be found in *The Bible Unearthed* by Israel Finkelstein and Neil Asher Silberman (Simon and Schuster, New York, 334 pp., 2001).

2 THE UNIVERSE

1. Galileo, G. Letter on Sunspots, quoted in Discoveries and Opinions of Galileo (Drake, S. Anchor Books, NY, p. 97, 1957).
2. Shakespeare, W. *As You Like It*, Act III Scene 2, 1599.
3. This classification of stars, dating from the work of Annie Jump Cannon at around 1920, was extended but not basically changed by Morgan and Keenan, and is also known as the MK classification. For a summary, see Garrison, R. Classification of Stellar Spectra, in *Encyclopedia of Astronomy and Astrophysics*, Institute of Physics Publishing, UK, 2001.
4. See for example, McKee, C. F. and Ostriker, E. C. Theory of star formation. AARA, Vol. **45**, pp. 565–687, 2007.
5. The Chemical Composition of the Sun, Asplund M. *et al.* ARAA Vol. **47**, pp. 481–522, 2009.
6. Fabrycky, D. C. Non-Keplerian dynamics of exoplanets, in *Exoplanets* (Editor S. Seager), University of Arizona Press, p. 226, 2010.
7. See Roberge, A. and Kamp, I. Protoplanetary and debris disks, in *Exoplanets* (Editor S. Seager), University of Arizona Press, p. 285, 2010, for an extended discussion.

8. Mittlefehldt, D. W. *et al.* in *Planetary Materials* (Ed. J. J. Papike), Reviews in Mineralogy, Vol. **36**, pp. 4–142 (Miner. Soc. Amer. Washington DC), 1998.
9. For those interested, in further details, a basic reference is Yin, Qing-zhu, From dust to planets, in *Chondrites and the Protoplanetary Disk* (ASP Conference Series Vol. **341**, pp. 632–642, 2005).
10. Jonson, Ben, *The Alchemist*, Act II, Scene 3, 1610.

3 FORMING PLANETS

1. Soter, S. Are planetary systems filled to capacity? *Amer. Scientist* Vol. **95**, p. 421, 2007.
2. However, the fate of the unfortunate beast may be a fable. The story of the elephant, although often told, is not mentioned in the three standard works on the investment of the city: Goure L., *The Siege of Leningrad*, Stanford University Press, 363 pp., 1962; Pavlov D. V. *Leningrad 1941*, University of Chicago, 186 pp., 1965; Salisbury H. E. *The 900 Days: The Siege of Leningrad*, Harper and Row, New York, 635 pp., 1969.
3. Kelvin, Lord (Thomson, W.) On the Origin of the Sun's Heat, *Popular Lectures and Addresses* Vol. **1**, Macmillan, London, 2nd edition, pp. 421–422, 1891.
4. Exoplanets (Ed. S. Seager), Arizona University Press, p. xvii, 2010.
5. Thommes, E. B. *et al.* Gas disks to gas giants: simulating the birth of planetary systems. *Science* Vol. **321**, p. 814, 2008.
6. Jupiter (eds. F. Bagenal *et al.*), Cambridge University Press, 719 pp., 2004
7. Lubow, S. H. and Ida, S. Planet migration, in *Exoplanets* (Ed. S. Seager), Arizona University Press, p. 347, 2010.
8. Tsiganis, K. *et al.* Origin of the orbital architecture of the giant planets in the solar system. *Nature* Vol. **435**, pp. 459–461, 2005.
9. Grieve, R. A. F. in *Meteorites: Flux with Time and Impact Effects* (eds. M. M. Grady *et al.*), *Geol. Soc. London Spec. Pub.* 140, p. 120, 1998.
10. Science, Vol. **334**, pp. 1616–1617, 2011.
11. Kramers, J. D. Hierarchical Earth accretion and the Hadean Eon. *Jour. Geol. Soc. London*, Vol. **164**, p.11, 2007, who comments that the Hf-W system does not date any event.
12. Goldschmidt, V. M. *Geochemistry*, Oxford University Press, Oxford, UK, 730 pp., 1954.

13. Lissauer, J. J. Chaotic motion in the solar system. *Rev. Modern Phys.* Vol. **71**, pp. 835–845, 1999.
14. See *Planetary Crusts* by S. R. Taylor and S. M. McLennan, Cambridge University Press, 378 pp., 2009.
15. Gillispie, C. C. *Genesis and Geology*, Harvard University Press, p. 127, 1951.
16. Weissman, P. R. in *Origin and Evolution of Planetary and Satellite Atmospheres* (Editors S. K. Atreya *et al.*), University of Arizona Press, Tucson, p. 241, 1989.
17. Gomes, R. *et al.* Origin of the cataclysmic Late Heavy Bombardment period of the terrestrial planets. *Nature* Vol. **435**, pp. 466–469, 2005.

4 THE EXOPLANETS

1. de Pater, I and Lissauer, J. J. *Planetary Science*, Cambridge University Press, p. 509, 2010.
2. Seager, S. and Lissauer, J. J. in *Exoplanets* (Ed. S. Seager), Arizona University Press, p. 4, 2010.
3. Meadows, V. and Seager, S. in *Exoplanets* (Ed. S. Seager), Arizona University Press, p. 441, 2010.
4. Exhaustive treatments can be found in *Exoplanets* (Ed. S. Seager), Arizona University Press, pp. 27–53, 2010 and in *The Exoplanet Handbook*, by Michael Perryman, Cambridge University Press, Chap. 2, 2011.
5. Bean, Jacob, *Nature* Vol. **478**, p. 41, 2011.
6. Jayawardhana, Ray, *Strange New Worlds*, Princeton University Press, 255 pp., 2011.
7. Mordasini, C. *et al. AA*, Vol. **501**, p. 1161, 2009.
8. Marcy, G. *et al. Prog. Theor. Phys. Supp.* Vol. **158**, 2005.
9. Cassan, A. *et al. Nature* Vol. **481**, pp. 167–169, 2012, Sumi, T. *et al.*, *Nature* Vol. **473**, pp. 349–352, 2011.
10. Raymond, S. N. *et al.* The debris disk-planet connection. IAU Vol. **276**: *Astrophysics of Planetary Systems*, 2011.
11. See Lubow, S. H. and Ida, S. Planet migration, in *Exoplanets* (Ed. S. Seager), Arizona University Press, p. 347, 2010.
12. Attributed to Doug Lin, University of California, Santa Cruz, a leading planetary theorist.
13. Baumann, P. *et al.*, *AA* Vol. **519**, p. A 87, 2010.
14. *Science* Vol. **286**, pp. 2239–2243, Dec 17, 1999.

15. Lovis, C. *et al.*, *AA*, Vol. **528**, p. 112, p. 201, 2011. HD are the initials of Henry Draper, who in the late nineteenth century funded a catalog of stars.
16. Lovis, C. *AA* Vol. **528**, p. 14, 2011.
17. Fortney, J. in *Exoplanets* (Ed. S. Seager), Arizona University Press, p. 414, 2010.
18. Correias, A. and Laskar, J. in *Exoplanets* (Ed. S. Seager), Arizona University Press, p. 259, 2010.
19. Sotin, C. in *Exoplanets* (Ed. S. Seager), Arizona University Press, p. 385, 2010.
20. Sotin, C. *et al.* in *Exoplanets* (Ed. S. Seager), Arizona University Press, p. 375, 2010.
21. Lascar, J. *Nature*, Vol. **361**, p. 615, 1993.
22. Those interested in this problem should consult Elser, S. *et al. Icarus*, Vol. **214**, p. 357, 2011.
23. See "Making other Earths: dynamical simulations of terrestrial planet formation and water delivery". Raymond, S. N. *et al.*, *Icarus*, Vol. **168**, pp. 1–17, 2004 and Lammer, H. *et al. AAR* Vol. **17**, p. 226, 2009.
24. Bond, J. C. *et al.* The compositional diversity of extra-solar terrestrial planets. I. In situ simulations. *ApJ*. Vol. **715**, pp. 1050–1070, 2010.
25. Compare O'Neill, C. and Lenardic, A. *GRL*, Vol. **34**, p. L19204, 2007 and Valencia, D. *et al.*, *ApJ*. L Vol. **670**, pp. L45–L48, 2007, for completely opposite opinions. See also Lammer, H. *et al. AAR*, Vol. **17**, p. 107, 2009.
26. Campbell, I. H. and Taylor, S. R. No water, no granites – No oceans, no continents. *GRL* Vol. **10**, pp. 1061–1064, 1983.
27. Cuntz, M. *et al.* On the plausibility of Earth-type habitable planets around 47 UMa. *Icarus* Vol. **162**, p. 217, 2003.
28. Basalla, G. *Civilized Life in the Universe*, Oxford University Press, p. 201, 2006.

5 OUR SOLAR SYSTEM

1. Mitchell, O. (1869) quoted by Brush, S. G. *A History of Modern Planetary Physics*, Vol. **1**, Cambridge University Press, p. 93, 1996.
2. See under Sources in the Appendices.
3. For an historical discussion, see Nieto, M. M. *The Titius – Bode Rule of Planetary Distances: Its History and Theory*, Pergamon, Elmsford, NY, 161 pp., 1972.

4. Gomes, R. *et al.* Origin of the cataclysmic Late Heavy Bombardment period of the terrestrial planets. *Nature,* Vol. **435**, pp. 466–469, 2005; Tsiganis, K. *et al.* Origin of the orbital architecture of the giant planets in the solar system, *Nature,* Vol. **435**, pp. 459–461, 2005; Morbidelli, A. *et al.* Chaotic capture of Jupiter's Trojan asteroids in the early solar system, *Nature* Vol. **435**, pp. 462–465, 2005.
5. Cooper, Henry S. F. *The New Yorker,* June 18, p. 73, 1990.
6. See The Origin and Evolution of Titan by J. Lunine *et al.* in *Titan from Cassini-Huygens* (Ed. R. Brown *et al.*), Springer, Netherlands, pp. 35–59, 2010 and Lammer, H. *et al. AAR,* Vol. **17**, p. 238, 2009.
7. Laplace, P. S. *The System of the World,* Vol. 1, Book I (J. Pond, Trans), R. Phillips, London, p. 97, 1809.
8. *Science,* Vol. **334**, pp. 1616–1617, 2011.
9. An authoritative and very entertaining account of the fall and subsequent history of this meteorite is given by Ursula Marvin, in *Meteoritics,* Vol. **27**, pp. 28–72, 1992.
10. Among the host of publications on meteorites, one of the best is by Alex Bevan and John de Laeter: *Meteorites: A Journey Through Space and Time,* University of NSW Press, Sydney, Australia, 215 pp., 2002.
11. For data from the Messenger mission to Mercury, see *Science,* Vol. **333**, pp. 1847–1868, September 30, 2011.
12. A good place to begin among the wealth of information about Mars is *The Martian Surface* (editor, Jim Bell), Cambridge University Press, 636 pp., 2008.
13. Comment by Richard Kerr in *Science Now,* Nov. 2, 2011. See also the important paper by Bethany Ehlmann *et al.* Subsurface water and clay mineral formation during the early history of Mars. *Nature,* Vol. **479**, pp. 53–60, 2011.
14. Toon, B. *et al.* The formation of martian river valleys by impacts. *Ann. Rev. Earth Planet. Sci.* Vol. **38**, pp. 303–322, 2010.

6 EARTH AND MOON

1. Seager, S. and Lissauer, J. J. in *Exoplanets* (Editor: Sara Seager), Cambridge University Press, p. 4, 2010.
2. Updike, J. *The New Yorker,* p.107, June 3, 1991.
3. Grotzinger, J. P. *et al. Nature Geoscience* Vol. **4**, pp. 285–292, 2011.

4. Taylor, S. R. and McLennan, S.M. *Planetary Crusts*, Cambridge University Press, p. 7, 2011.
5. Preston Cloud, *Oasis in Space*, Norton, New York, 508 pp., 1988, gives the best general account of the geological history of the Earth.
6. Cloud, Preston (1988), *Oasis in Space*, Norton, New York, p. 267.
7. Hoffman, P. F. *et al.* A Neoproterozoic Snowball Earth. *Science* Vol. **381**, pp. 1342–1346, 1998.
8. Marshall, C. R. Explaining the Cambrian "explosion" of animals. *Ann. Rev. Earth Planet. Sci.* Vol. **34**, pp. 55–384, 2006.
9. See *Planetary Crusts* (S. R. Taylor and S. M. McLennan, Cambridge University Press, 378 pp., 2009) for an extended discussion.
10. Currently accepted models for delivering water to the Earth can be found in Morbidelli, A. *et al. Meteoritics and Planetary Science*, Vol. **35**, pp. 1309–1320, 2000 and Lunine, J. L. in *Meteorites and the Early Solar System* (Editors: D. S. Lauretta and H. Y. McSween), Arizona University Press, pp. 309–319, 2006.
11. Tennyson, Alfred Lord, *Locksley Hall*, 1835.
12. Kasting, J. F. and Siefert, L. L. *Science*, Vol. **296**, p. 1066, 2002.
13. P. Z. Myers in a review of Paul Davies, *The Eerie Silence.*
14. Gould, S. J. *Wonderful Life: The Burgess Shale and the Nature of History*, Penguin, London, 347 pp., 1991.
15. Burger, W. C. *Flowers, How they Changed the World.* Prometheus Books 337, pp., 2006.
16. Pollard, W. G. *Amer. Scientist* Vol. **67**, p. 654, 1979.
17. For those interested in pursuing these fascinating topics further, a good place to begin are the papers by Mark Maslin and Beth Christensen "Tectonics, orbital forcing, global climate change, and human evolution in Africa", *Journal of Human Evolution* Vol. **53**, pp. 443–464, 2007 and Rosenberg, Karen R. and Trevathan, Wenda, Birth, obstetrics and human evolution. *BJOG*, Vol. **109**, 1199–1206, 2002.
18. Conway Morris, S. *Life's Solution*, Cambridge University Press, 464 pp., 2003.
19. Ralling, C. *The Voyage of Charles Darwin (His Autobiographical Writings)*, Aerial Books, London p. 73, 1982.
20. Barnosky, A. D. *et al.* Has the Earth's sixth mass extinction already arrived? *Nature* Vol. **471**, pp. 51–57, 2011.
21. Scott, R. F. *Earth Science Reviews* Vol. **13**, p. 379, 1977.

22. Laplace, P. S. *The System of the World*, Vol. **1**, Book IV (J. Pond, Trans), R. Phillips, London, p. 94, 1809.
23. Dawkins, R. *The Blind Watchmaker*, Norton, New York, p. 6, 1987.
24. The Rosetta Stone was discovered in 1799 during the French invasion led by Napoleon, near Rosetta, now Rashid, in Egypt. It is a tablet of black basalt, on which are listed benefactions by Ptolemy V Epiphanes (205–180 BCE). Priests at Memphis had inscribed this tablet in two Egyptian scripts (hieroglyphs and demotic, a script related to hieroglyphs) and crucially also in Greek. This discovery enabled the ancient Egyptian picture language (hieroglyphs) to be translated, a task accomplished mainly by J.F. Champollion by 1822.
25. Detailed accounts of the results from the Apollo Missions can be found in *New Views of the Moon* (eds. B. Jolliff *et al.*) *Rev. Min. Geochem.* Vol. **60**, 2006.
26. For a detailed account of the giant impact model, see Canup, R. M. Dynamics of lunar formation. *AARA* Vol. **42**, pp. 441–475, 2004.
27. Lascar, J. *et al. Nature*, Vol. **361**, p. 615, 1993.
28. Huygens, C. *The Celestial Worlds Discovered*, T. Childe, London, p. 131, 1698.
29. Sagan, C. *Space Life Sci.* Vol. **3**, p. 484, 1972. This quotation provides yet another example of the use of double negatives that are popular in astrobiology.

7 PERSPECTIVES

1. Ralling, C. *The Voyage of Charles Darwin (His Autobiographical Writings)*, Aerial Books, London p. 130, 1982.
2. Those interested in this topic may make a comparatively sober beginning by consulting the following excellent sources: R. A. S. Hennessey, *Worlds Without End* (Tempus, UK, 1999), George Basalla, *Civilized Life in the Universe* (Oxford University Press, UK, 2005), Michael J. Crowe, *The Extraterrestrial Life Debate* (University of Notre Dame Press, Indiana, 2008), Steven J. Dick, *Life on Other Worlds* (Cambridge University Press, UK, 1998) and *The Biological Universe* (Cambridge University Press, 1996) and Mark Brake, On the plurality of inhabited worlds: a brief history of extra-terrestrialism. *International J. Astrobiology* Vol. **5**, pp. 99–107, 2006.
3. Fauser, W. (1982) *Die Werke des Albertus Magnus*, Aschendorff, Munster, Vol. **V**, Pt 1. (Trans. S. J. Dick).

4. See, for example, Fontenelle, B. de *Conversations on the plurality of worlds*, Withy, London, 1686, English trans. 1760; Derham, W. *Astro-Theology, W*. Innys, London,1715; Whewell, W. *On the Plurality of Worlds*, Parker, London, 1853; Lovejoy, A. O. *The Great Chain of Being*, Harvard University Press, Cambridge MA, 382 pp., 1936; Sagan, C. and Shklovskii, I. S. *Intelligent Life in the Universe*, Dell, New York, 509 pp., 1966; Sagan, C. *Boca's Brain*, Random House, New York, 347 pp., 1979.
5. The celebrated Drake Equation allows one to calculate the number of civilizations in the universe. However, it involves so many unknowable probabilities multiplied together that it can generate any number that the inquirer fancies. It thus remains an intellectual curiosity, much like the concept of Gaia, which proposes that life forms a self-regulating system that controls the environment on this planet.
6. Kasting, J. *How to Find a Habitable Planet*, Princeton University Press, 2010.
7. Fayé, H. *Sur l'origine due Monde* (2nd edn), Gauthier-Villars et fils, Paris, pp. 299–300, 1885.
8. Kevin Zahnle, NASA Ames, personal communication.
9. Lammer, H. *et al.* What makes a planet habitable? *AAR* Vol. **17**, pp. 181–249, 2009. This lengthy paper with 17 coauthors reads like the deliberations of a committee.
10. Nevertheless a paper with 32 co-authors, including seven from the SETI Institute, concludes, in a statement circumscribed by caveats "that M dwarf stars may indeed be viable hosts for planets on which the origin and evolution of life can occur" (Tarter, J. *et al. Astrobiology*, Vol. **7**, pp. 30–65, 2007). Another paper makes use of that useful device, the double negative, by concluding, "that no known phenomenon completely precludes the habitability of terrestrial planets orbiting cool stars". Barnes, R. *et al.* in *Cool Stars XVI*, ASP Conference Series 448, 2010.
11. Stephen Greenblatt discussing Lucretius in *The New Yorker*, p. 30, August 8, 2011.
12. The concept of "Élan vital" is due to Henri Bergson (1859–1941) but was dismissed by Julian Huxley (1887–1975) as no better than "explaining the operation of a railway engine by its élan locomotif".
13. Crick, F. *The Astonishing Hypothesis: The Scientific Search for the Soul*, Simon and Schuster, New York, 317 pp., 1994.

14. See Andrew J. Watson, Implications of an anthropic model of evolution for emergence of complex life and intelligence, in *Astrobiology* Vol. **8**, pp. 175–185, 2008 for a theoretical basis for evolution.
15. Marc Kaufman, *First Contact: Scientific Breakthroughs in the Hunt for Life Beyond Earth*, Simon and Schuster, 224 pp., 2011.
16. Mayr, E. in *Perspectives in Biology and Medicine*, Vol. **37**, pp. 150–154, 1994; see also *The Planetary Report*, Vol. **16** (3), p. 6, 1996.
17. Frank J. Tipler. A brief history of the extraterrestrial intelligence concept. *QJRAS* Vol. **22**, pp. 133–145, 1981.
18. De Pater, I. and Lissauer, J. J. *Planetary Sciences*, Cambridge University Press, p. 587, 2010.
19. See Barsalla, G. *Civilized Life in the Universe*, pp. 111–112, Oxford University Press, 2006, for comments about the use of double negatives in astrobiology.
20. Ferris, T. in *The New Yorker*, p. 31, April 14, 1997.
21. Edward Gibbon, *The Decline and Fall of the Roman Empire* (abridged, D. M. Low), Chatto and Windus, London, Chap. 15, 1978.
22. Cromer, A. *Uncommon Sense*, Oxford University Press, 240 pp., 1993; Huff, T. E. *Intellectual Curiosity and the Scientific Revolution*, Cambridge University Press, 368 pp., 2011.
23. Peter Medawar in his celebrated review of Chardin's "The Phenomenon of Man" claims that "before deceiving others, he (Chardin) has taken great pains to deceive himself". See Medawar, Peter, *Pluto's Republic*, Oxford University Press, p. 242, 1984. Self-deception is a well-known psychological phenomenon. See Trivers, R. *Deceit and Self-Deception*, Allen Lane, 2011.
24. Jerry Coyne, review of *Can a Darwinian be a Christian?* by Michael Ruse in *London Review of Books*, Vol. **24** (9), pp. 23–24, 2002.
25. Davies, P. C. W. *The Mind of God*, Simon and Schuster, New York, p. 232, 1992.
26. Monod, J. *Chance and Necessity*, Collins Fontana, London, pp. 154 and 167, 1974.
27. Steven Weinberg, in *The New York Review of Books*, Oct 21, 1999.
28. See Beckmann, P. *A History of Pi* (π). Golem Press, Boulder, CO., p. 170, 1971. Curiously, a value of 3.0 for π occurs in the description of the building of the Temple of Solomon in Jerusalem in *I Kings Chapter 7, verse 23* and in *II Chronicles Chapter 4, verse 2*. The often

repeated story that a state legislature (variously Illinois, Indiana or Massachusetts) in America were sufficiently impressed by the biblical value of π that they attempted in the nineteenth century to pass a law stating that the value of π was 3.0, appears to be without foundation.

29. Dawkins, R. *The Selfish Gene*, Oxford University Press, p. 325, 1989.
30. Shakespeare, W. *The Tempest*, Act I Scene 2, 1611.
31. J. D. Barrow and F. J. Tipler discuss the anthropic principle at length in *The Anthropic Cosmological Principle*, Oxford University Press, 706 pp., 1986.
32. Howard A. Smith, Alone in the Universe, Talk at Amer. Assoc. Advance. Science, Washington, DC, Meeting February 20, 2011.
33. Tipler, F. J. *QJRAS* Vol. 22, p. 133, 1981.
34. Shakespeare, W. *Macbeth*, Act 1 Scene 7, 1606.
35. Basalla, G. *Civilized Life in the Universe*, Oxford University Press, 233 pp., 2006.

Index